ELECTRICAL INSTALLATION COMPETENCES

Part I Studies: Science

by the same author

Electrical Installation Technology 1
Electrical Installation Technology 2
Questions and Answers in Electrical Installation Technology
Electrical Installation Competences Part 1 Studies: Practical
Electrical Installation Competences Part 1 Studies: Science
Electrical Installation Competences Part 1 Studies: Theory
Electrical Installation Competences Part 2 Studies: Practical
Electrical Installation Competences Part 2 Studies: Science

ELECTRICAL INSTALLATION COMPETENCES

Part I Studies: Science

Maurice Lewis
BEd (Hons), FIEIE

Stanley Thornes (Publishers) Ltd

First published in 1991 by
Stanley Thornes (Publishers) Ltd
Ellenborough House
Wellington Street
CHELTENHAM GL50 1YD

Reprinted 1993
Reprinted 1994

British Library Cataloguing in Publication Data

Lewis, Maurice
Electrical installation competences: Part I.
Science.
I. Title
621

ISBN 0-7487-0591-0

Cover photograph by courtesy of MEGGER INSTRUMENTS LIMITED

Typeset by Tech-Set, Gateshead, Tyne & Wear.
Printed and bound in Great Britain at The Bath Press, Avon.

Contents

Preface vii

1 Terminology 1

Definitions of SI and derived units; metric system of measurement; powers of a quantity (indices); useful metric weights and measures; transposition of formulae; mensuration; graphs and charts; ratio, proportion and percentage; BS3939 circuit symbols; miscellaneous terms and symbols; exercise 1.

2 Mechanical science 15

Mass, force and weight; Newton's laws of motion; exercise 2.1; vector quantities; equilibrium; principle of moments; density; exercise 2.2; heat, temperature and expansion; gas laws; work, energy and power; simple machines; exercise 2.3.

3 Electrical science 30

Basic circuit theory; Ohm's Law; resistance factors; resistor connections; resistors in series; resistors in parallel; energy and power; heating effects of electric current; power factor; efficiency; magnetic effects of electric current; simple loop generators; transformers; chemical effects of electric current; secondary cells; exercise 3.

4 Electronics 54

Electronic components; resistors; capacitors; inductors; semiconductor devices: diode; zener diode; light emitting diodes; photodiode; transistor; thyristor; diacs and triacs; thermistor; integrated circuits (ICs); alternating current waveforms; exercise 4.

Appendix 1 Multiple choice questions 71

Appendix 2 Short-answer questions 80

Appendix 3 Answers 84

Acknowledgements

Grateful acknowledgements are made to Mr C. R. Shotbolt and Mr S. W. Davis for their review of chapters and to Katherine Pate, editor at Stanley Thornes, for her management of the book from beginning to end.

Preface

This is the first of three electrical installation books at Part I level, which cover important topics in the new City and Guilds 236–7 syllabus scheme reflecting competence based learning.

This book covers mechanical and electrical science and also includes basic electronics, which has become a rapidly changing technology over the years, moving closer and closer to electrical installation work.

As an introduction, Chapter 1 concentrates on *terminology* to provide you with an understanding of electrotechnical terms and a mathematical background to support your reading of the other chapters.

Chapter 2, *Mechanical science*, attempts to give you an awareness of the many different principles and laws associated with the physical side of your work: the simple lever; equilibrium; centre of gravity; linear expansion, etc. These are very important topics that frequently apply to practical situations where you will have to offer your support and skills to overcome work problems.

Chapter 3, *Electrical science*, commences with basic circuit theory, and develops your awareness of the common effects of electric current such as its heating effect; magnetic effect and chemical effect. Other topics such as power factor and efficiency are also discussed.

Chapter 4 concentrates on *basic electronics*, introducing you to electronic components and semiconductor devices as well as a.c. theory and alternating and digital waveforms.

Multiple choice questions of the type you will be required to complete in your examinations are given in Appendix 1; Appendix 2 contains *short-answer questions*. The *answers* to all these questions and questions in the exercises in the chapters are given in Appendix 3.

The other books in this series, *Part I Studies: Practical* and *Part I Studies: Theory* are major reference books for the course and should be read in order to fully understand the topics covered. Other students studying City and Guilds Courses 185, 201 and 232 will find these three books of immense value.

Maurice Lewis

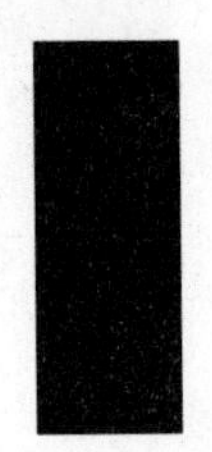

References

In order for you to extend your knowledge on topics in this book, you should make reference to *Part I Studies: Practical* and *Part I Studies: Theory* of this series.

Topic	*Science Chapter*	*Practical Chapter*	*Theory Chapter*
Study of electrical industry			1
Safe practices		1	1
Basic work preparation	2	1	2
Wiring systems		1, 2	3
Earthing equipment		2	3
Inspection and testing		2	3
Circuit fault finding		2	3
Circuit principles	2, 3, 4	4	
Basic electronic wiring		3	4

Health and Safety at Work Act, 1974, HMSO
SI 1989/2209 *The Construction (Head Protection) Regulations*, 1989, HMSO
SI 1988/1057 *The Electricity Supply Regulations*, 1988, HMSO
SI 1989/635 *Electricity at Work Regulations*, 1989, HMSO
Institution of Electrical Engineers Regulations for electrical installations: sixteenth edition, 1991, Stevenage, Herts

The following guidance notes are obtainable from the Health and Safety Executive, HMSO:

GS 6 *Avoidance of danger from overhead electrical lines*, 1980
GS 24 *Electricity on construction sites*, 1983
GS 27 *Electricity on construction sites*, 1984
GS 33 *Avoiding danger from buried electricity cables*, 1985
GS 34 *Electrical safety in departments of electrical engineering*, 1986
GS 37 *Flexible leads, plugs, sockets, etc*, 1985
GS 38 *Electrical test equipment for use by electricians*, 1986
GS 44 *Electrical working practices* (in preparation)
PM 32 *Safe use of portable electrical apparatus (electrical safety)*, 1983
PM 38 *Selection and use of electric handlamps*, 1984
HS(G) 13 *Electrical testing: safety in electrical testing*, 1980
HS(G) 38 *Lighting at work*, 1987
HS(R) 7 *A guidance to the Safety Signs Regulations*, 1980
HS(R) 25 *Memorandum of Guidance on the Electricity at Work Regulations*, 1989
HS(R) 23 *Guide to the Reporting of Injuries, Diseases and Dangerous Occurrences Regulations*, 1985, 1986.

Terminology

1

Objectives

After reading this chapter you should be able to:

- *state the basic principles, units and derived units of the SI system.*
- *recognise and use the basic symbols for quantities and their units in the SI system.*
- *name and give the symbol and factor appropriate to multiples and submultiples of SI units.*
- *apply the metric system of measurement to practical situations.*
- *perform basic calculations including the transposition and use of basic formulae related to SI and derived SI units.*
- *perform basic calculations to find length, area and volume of figures.*
- *recognise and draw a number of common BS3939 electrical circuit symbols.*

Definitions of SI Units

The UK and many other European countries use the international system of units known as SI Units. This system, which developed from the metric system, comprises seven base units and two supplementary units as shown in Table 1.

Table 1 SI Units

Quantity	*Unit name*	*Unit symbol*
Length	metre	m
Mass	kilogram	kg
Time	second	s
Electric current	ampere	A
Thermodynamic temperature	kelvin	K
Amount of substance	mole	mol
Luminous intensity	candela	cd
Plane angle	radian	rad
Solid angle	steradian	sr

The quantities used in the electrical installation course are length (l), mass (m), time (t), current (I), temperature (T) and luminous intensity (I). These units are defined as follows.

Metre: the unit of length defined by the path travelled by light in a vacuum during a time interval of 1/299 792 458 of a second.

Kilogram: the unit of mass of a prototype block of platinum preserved at the International Bureau of Weights and Measures.

Second: the unit of time that is the duration of 9 192 631 770 periods of radiation.

Ampere: the unit of current, which if maintained in two straight parallel conductors of infinite length and of negligible circular cross section, placed 1 metre apart in a vacuum, would produce between them a force equal to 2×10^{-7} newtons per metre of length.

Kelvin: the fraction 1/273.16 of the thermodynamic temperature of the triple point of water. For practical purposes the degree Celsius scale is used. Both have identical intervals, i.e. 1 K = 1 °C, but the Kelvin scale begins at absolute zero, the coldest temperature possible where bodies possess no thermal energy, −273 °C. This means that freezing point (0 °C) is 273 K and boiling point (100 °C) is 373 K.

Candela: the unit of luminous intensity, perpendicular to a surface of 1/600 000 square metres of a black body at the temperature of freezing platinum. For practical purposes it is a measure of light output from a light source in a given direction.

Definitions of derived SI quantities and other electrotechnical terms

From the basic SI Units come numerous derived units. For example, a coulomb is the unit of charge derived from the product of current (A) and time (s). The following terms are frequently used in electrical/mechanical engineering.

Area: the surface enclosed by the sides of a two-dimensional shape (i.e. product of two lengths); measured in square metres (m^2).

Capacitance: the property of a capacitor to store an electric charge.

Charge: the excess of positive or negative electricity on a body, in space, or passing at a point in an electric circuit during a given time.

Density: the mass per unit volume of a substance; measured in kilograms per cubic metre (kg/m^3).

Electromotive force: the force necessary to cause the movement of charges.

Energy: the capacity or ability of matter or radiation to do work.

Force: the cause of mechanical displacement or motion, or the effect it produces.

Frequency: describes the number of repetitive cycles that occur in one second of an a.c. waveform. The duration of one cycle is the **periodic time.** The public supply in the UK has a frequency of 50 Hz.

Illuminance: the amount of light in lumens falling on a unit area of one square metre.

Impedance: the ratio of voltage and current in r.m.s. terms for a.c. quantities.

Inductance (self): the property of an inductor to produce a magnetic field when carrying current.

Luminous flux: the capacity of radiant energy to produce light.

Magnetic flux: the phenomenon associated with invisible 'lines of force' in the neighbourhood of magnets and electric currents.

Magnetic flux density: the quantity of magnetic flux spread over a given area.

Magnetomotive force: the force required to cause magnetic flux to flow.

Permeability: the ratio of magnetic flux density to magnetising force.

Potential difference: the cause of movement of electric charge from one point to another.

Power: the dissipation of energy, found by the product of voltage and current.

Table 2 Commonly used symbols and their units

Quantity	*Symbol*	*Unit name*	*Unit symbol*
area	*A*	square metre	m^2
capacitance	*C*	farad	F
charge	*Q*	coulomb	C
current	*I*	ampere	A
density	ρ	mass/unit volume	kg/m^3
electromotive force	*E*	volt	V
energy	*W*	joule	J
force	*F, f*	newton	N
frequency	*f*	hertz	Hz
illuminance	*E*	lux	lx
impedance	*Z*	ohm	Ω
inductance (self)	*L*	henry	H
luminous flux	*F*	lumen	lm
magnetic flux	Φ	weber	Wb
magnetic flux density	*B*	tesla	T
magnetomotive force	*F*	ampere-turn	A, At
permeability	μ	henry per metre	H/m
potential difference	*V*	volt	V
power	*P*	watt	W
reactance	*X*	ohm	Ω
reluctance	*S*	ampere per weber	A/Wb
resistance	*R*	ohm	Ω
resistivity	ρ	ohm metre	Ω m
volume	*V*	cubic metre	m^3

Reactance: the property of inductors and capacitors to resist the flow of alternating current.

Reluctance: the ratio of the magnetic force acting round a magnetic circuit to the resulting magnetic flux.

Resistance: the property of a resistor to resist the flow of charge through it.

Resistivity: the resistance measured between the opposite faces of a unit cube of given material.

Volume: the three-dimensional quantity of a container or solid; measured in cubic metres (m^3).

Note You should be aware that some quantities, such as current and luminous intensity; force and magnetomotive force; charge and quantity of a substance, have the same symbol. Table 2 on page 2 gives the symbols and units for the quantities defined above. Notice that the quantity symbols are in italic (e.g. C, Q). Try to present any formulae you write in this form.

Metric system of measurement

One advantage of the metric system is the ease with which you can use prefix symbols to replace multiples and sub-multiples of various units. Table 3 shows some of the prefixes commonly used in electrical work, together with their meanings and multiplying factors. Some examples of converting lengthy units into metric prefix units are given after the table.

Table 3 Metric prefixes

Prefix	*Symbol*	*Multiplying factor*	*Power/ index*
tera	T	1 000 000 000 000	(10^{12})
giga	G	1 000 000 000	(10^{9})
mega	M	1 000 000	(10^{6})
kilo	k	1 000	(10^{3})
hecto	h	100	(10^{2})
deca	da	10	(10^{1})
unity		1	(10^{0})
deci	d	0.1	(10^{-1})
centi	c	0.01	(10^{-2})
milli	m	0.001	(10^{-3})
micro	μ	0.000 001	(10^{-6})
nano	n	0.000 000 001	(10^{-9})
pico	p	0.000 000 000 001	(10^{-12})

Note 1 Mg = 1000 kg = 1 tonne (t).

Q

Example 1.1

Convert the following:

a) 10 500 000 joules to megajoules

b) 0.5 volts to millivolts

c) 20 202 watts to kilowatts

d) 600 milliamperes to amperes

e) 0.000 003 farads to microfarads

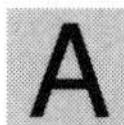

Solution

a)
$$1\,000\,000 \text{ J} = 1 \text{ MJ}$$
$$10\,500\,000 \text{ J} = 10.5 \text{ MJ}$$

b)
$$1 \text{ V} = 1000 \text{ mV}$$
$$0.5 \text{ V} = 0.5 \times 1000 \text{ mV}$$
$$= 500 \text{ mV}$$

c)
$$1000 \text{ W} = 1 \text{ kW}$$
$$20\,202 \text{ W} = 20.202 \text{ kW}$$

d)
$$1000 \text{ mA} = 1 \text{ A}$$
$$600 \text{ mA} = 600/1000 \text{ A}$$
$$= 0.6 \text{ A}$$

e)
$$1 \text{ F} = 1\,000\,000\ \mu\text{F}$$
$$0.000\,003 \text{ F} = 0.000\,003 \times 1\,000\,000\ \mu\text{F}$$
$$= 3\ \mu\text{F}$$

Example 1.2

Express the following in more convenient units.

a) 6600 watthours

b) 0.75 litres

c) 778 000 ohms

d) 0.099 milliamperes

e) 50 000 kilograms

Solution

a) 6.6 kWh

b) 75 cl

c) 778 kΩ

d) 99 μA

e) 50 Mg

Powers of a quantity (indices)

From the metric prefixes you can see that quantities or numbers may have positive or negative powers. For example, the metric prefix 'M', meaning mega, has a positive index 6, i.e. 10^6. This should be read as 'ten to the power of six' and it indicates how many 10s have to be multiplied together to make 1 000 000. In the same way the metric prefix 'm', meaning milli, has a negative index −3 and this is written 10^{-3}. It is read as 'ten to the power of minus 3'. These negative powers save you writing down the quantity as a fraction, such as $1/10^3$ or 1/1000 (one thousandth). You will probably have noticed that Appendix 8 of the *IEE Wiring Regulations* incorporates time-current characteristics of protective devices and that both axes chosen for the graphs are labelled in powers of ten. It is important, therefore, that the laws of indices are fully understood. A range of these powers of ten has already been given, but remember also that $10^1 = 10$ and $10^0 = 1$.

In multiplication, the index of the product of two powers of the same quantity is the sum of their separate indices. For example,

a) $10^3 \times 10^6 = 10^{3+6} = 10^9$

b) $10^2 \times 10^{-3} = 10^{2-3} = 10^{-1}$

c) $10^6 \times 10^{-6} = 10^{6-6} = 10^0 = 1$

In division, the index of the divisor must be subtracted from the index of the dividend (numerator). For example,

a) $10^3 \div 10^6 = 10^{3-6} = 10^{-3}$

b) $10^2 \div 10^{-3} = 10^{2+3} = 10^5$

c) $10^6 \div 10^{-6} = 10^{6+6} = 10^{12}$

In instances where metric prefixes are not given, positive indices abbreviate high values and negative indices abbreviate very small values. For example:

a) the Supergrid system operates at 400×10^3 V (i.e. 400 kV).

b) the insulation resistance of a cable is measured to be $200 \times 10^6\ \Omega$ (i.e. 200 MΩ).

c) a capacitor is designed with a capacitance of 100×10^{-6} F (i.e. 100 μF).

d) an inductor is designed with an inductance of 0.5×10^{-3} H (i.e. 0.5 mH).

Resistor and capacitor colour codes frequently use metric prefixes. They are a form of identification used for telling the value of small circuit elements. The idea arose out of the difficulty in reading the components once they are mounted in position.

Table 4 Resistor colour codes

Colour	*Significant Figures*	*Multiplier*	*Tolerance*
silver	—	10^{-2}	±10%
gold	—	10^{-1}	±5%
black	0	1	—
brown	1	10	±1%
red	2	10^2	±2%
orange	3	10^3	—
yellow	4	10^4	—
green	5	10^5	±0.5%
blue	6	10^6	±0.25%
violet	7	10^7	±0.1%
grey	8	10^8	—
white	9	10^9	—
none	—	—	±20%

Table 4 shows the resistor colour codes used to overcome this problem. The colour bands represent significant figures, multipliers and tolerance and the first thing to observe is that the bands are to one end of the resistor and should always be read from that end. The first band tells us the value between 1 and 9, e.g. red is 2, and the second band also tells us the value between 1 and 9, e.g. violet is 7. The third band tells us the number of noughts to be added to the digits 2 and 7 and it is very important for it to be read correctly otherwise a gross error could be made. Here the orange band is 3, thus making 27 000 ohms (or 27 kΩ). The fourth band is the tolerance or range within which the manufacturer guarantees the value of the resistor. In this case it is coloured gold and this tells us that its value can be 5% more or 5% less than the stated value.

Another method of determining resistor values is using a letter and digit code. In this method the code letters replace the decimal point. The letter R

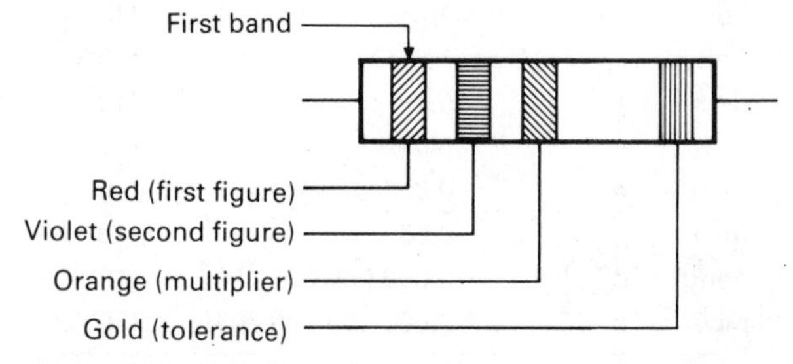

Figure 1.1 Resistor colour codes

is used as the decimal point and the letters K, M, G and T are used as multipliers for 10^3, 10^6, 10^9 and 10^{12}. For example a resistor of 0.27 Ω is coded R27 and one which is of 27.32 Ω is coded 27R32. A resistor of 100 Ω is coded 100R. In terms of other letters, a resistor of value 10 kΩ is coded 10K and one that has a value 10.33 kΩ is coded 10K33. A 1 MΩ resistor would be coded 1M0 and a 10 MΩ would be coded 10M. For tolerances, the letters B, C, D, F, G, J, K, M and N are used signifying ±0.1%, ±0.25%, ±0.5%, ±1%, ±2%, ±5%, ±10%, ±20% and ±30% respectively. If a 4.7 MΩ resistor had a tolerance of 20% it would be coded 4M7M.

Useful metric weights and measures

LINEAR MEASURE

10 millimetres (mm)	=	1 centimetre (cm)
10 cm	=	1 decimetre (dm)
10 dm	=	1 metre (m)
10 m	=	1 decametre (dam)
10 dam	=	1 hectometre (hm)
10 hm	=	1 kilometre (km)

SQUARE MEASURE

100 mm^2	=	1 cm^2
100 cm^2	=	1 dm^2
100 dm^2	=	1 m^2
100 m^2	=	1 dam^2
100 dam^2	=	1 hm^2
100 hm^2	=	1 km^2

CUBIC MEASURE

1000 mm^3	=	1 cm^3
1000 cm^3	=	1 dm^3
1000 dm^3	=	1 m^3

LIQUID MEASURE

10 centilitres (cl)	=	1 decilitre (dl)
10 dl	=	1 litre (l)
10 l	=	1 decalitre (dal)
10 dal	=	1 hectolitre (hl)

USEFUL EQUIVALENTS

1 litre of water	=	1 kilogram (kg)
1000 kg	=	1 tonne (t)
1000 l	=	1 m^3
1 unit of electricity	=	1 kilowatt hour (kWh)
1 kWh	=	3.6 megajoules (MJ)
1 joule (J)	=	1 newton metre (Nm)

Transposition of formulae

You will have noticed by now that there are a large number of formulae involved in electrical work. To solve a particular problem the formula often has to be rearranged to make the required term the subject. The first thing to appreciate in any formula is that the multiplication sign between term symbols is often omitted but it should always be inserted when the terms are replaced by numbers.

Let us consider the formula

$$Q = It$$

which is explained in more detail later. Each of the terms has a definite meaning and Q is found by multiplying together the product factors I and t. To make I the subject of the formula, t has to be moved. The way of dealing with this is to divide both sides of the formula by t.

Thus, if $$Q = It$$

then $$\frac{Q}{t} = \frac{It}{t}$$

It will be seen that t on the right-hand side can be cancelled out, leaving Q/t on the left-hand side and I on the right-hand side.

Thus $$\frac{Q}{t} = I$$

or $$I = \frac{Q}{t}$$

The same procedure can be repeated in order to find t. If there are more than two unknown product factors, just divide by those not required.

Let us look at another formula, $R = \rho l/a$. Say we require to make a the subject of the formula. The simplest method here is to turn the formula upside down so that it becomes:

$$\frac{1}{R} = \frac{a}{\rho l}$$

Now just multiply ρl on the top line of both sides of the formula and again we will see that the ρl on the right-hand side of the equal sign will cancel and disappear, leaving a on its own.

Thus $$\frac{\rho l}{R} = \frac{\rho l a}{\rho l}$$

and $$\frac{\rho l}{R} = a$$

or $$a = \frac{\rho l}{R}$$

It must be remembered that the transposing procedure requires *you* to make decisions but you must know the rules. Some of these are explained below.

Take a further example, such as $V = E - IR$. Say we had to find R. Here, we do not need to divide both sides by I at first. It is easier to make $-IR$ positive by taking it across to the other side of the equal sign, thus

$$V + IR = E$$

Now bring V across to the right-hand side where it becomes $-V$

$$IR = E - V$$

Applying the above procedure, divide both sides of the equal sign by I since we want to make R the subject of the formula.

Thus $$\frac{IR}{I} = \frac{(E - V)}{I}$$

therefore $$R = \frac{(E - V)}{I}$$

One final example to show is the formula using $Z = \sqrt{R^2 + X^2}$. Here we are dealing with a square root sign and the first job is to rearrange this formula in terms of Z^2.

For example, if $5 = \sqrt{25}$

then $5^2 = 25$

Hence $Z^2 = R^2 + X^2$

The square root vanishes. If X is required, then R^2 will become $-R^2$ on the left-hand side of the equal sign.

Thus $$Z^2 - R^2 = X^2$$

This can now be rearranged so that X is the subject of the formula and is found by square rooting the rest of the formula.

Therefore $$X = \sqrt{Z^2 - R^2}$$

The term R can also be found this way. It should be pointed out that both R and X will always be less than Z since the formula relates to Pythagoras' Theorem (as explained in 'Mensuration' below).

Q

Example 1.3

Transpose the following formulae:

a) $A = \pi r^2$ make r the subject of the formula

b) $X = 1/2\pi fC$ make f the subject of the formula

c) $P = V^2/R$ make V the subject of the formula

d) $R_1 = R_0(1 + \alpha t)$ make t the subject of the formula

e) $IR = E - Ir$ make I the subject of the formula

Solution

a) $r = \sqrt{\dfrac{A}{\pi}}$

b) $f = \dfrac{1}{2\pi XC}$

c) $V = \sqrt{PR}$

d) $t = \dfrac{R_1 - R_0}{R_0 \alpha}$

e) $I = \dfrac{E}{R + r}$

Mensuration

This section is concerned with the mathematical rules for finding lengths, areas, volumes, etc. of figures. First let us consider the measurement of angles. In Figure 1.2, let us suppose the straight line OX starts from the position OA and moves through the positions OB, OC, OD and back to OA so that it describes a circle about O in an anticlockwise direction. The line OX generates the angle AOX as it passes round to complete one revolution. In doing so the angle is divided into 360 equal parts

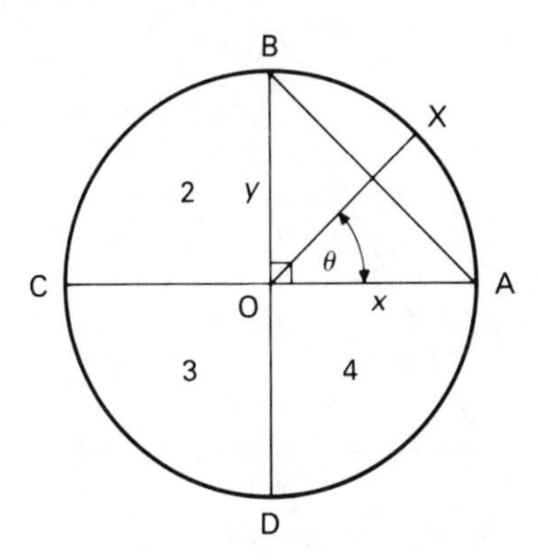

Figure 1.2 Measurement of angles

called degrees. One degree is divided into 60 minutes and one minute is divided into 60 seconds. It will be seen that the completed circle is divided into four equal parts called **quadrants**, each containing a **right angle**, i.e. an angle of 90°. If you measure angle AOX in the first quadrant with a protractor you will find it to be 45°.

Some further terms relating to the geometry of the circle are:

AO is the **radius**

AC is the **diameter**

AX is the **arc** created by the angle AOX. For the completed circle it is called the **circumference**.

AB is a **chord**. It divides the circle into two **segments**. (The **minor segment** is the smaller one.)

A **sector** is an area bounded by two radii and the arc between them (quadrants are sectors).

Within the first quadrant triangle AOB is a right-angled triangle and if you use a protractor on the other angles you will notice that they are both 45°. The sum of these angles is 180°. In fact the sum of the angles of any triangle is 180°.

Triangle AOB in Figure 1.2 represents a square cut in half and the diagonal line AB is the longest side of the triangle, called the **hypotenuse**. For any square the hypotenuse is always 1.414 ($\sqrt{2}$) times longer than the other sides. A simple method of finding an unknown side of a right-angled triangle is to use Pythagoras' Theorem, which states:

> In a right-angled triangle the square on the hypotenuse is equal to the sum of the squares on the other two sides.

Draw a right-angled triangle with sides measuring 3 cm, 4 cm and 5 cm and complete the squares as indicated by the theorem (see Figure 1.3). It will be seen that $AB^2 = AO^2 + BO^2$

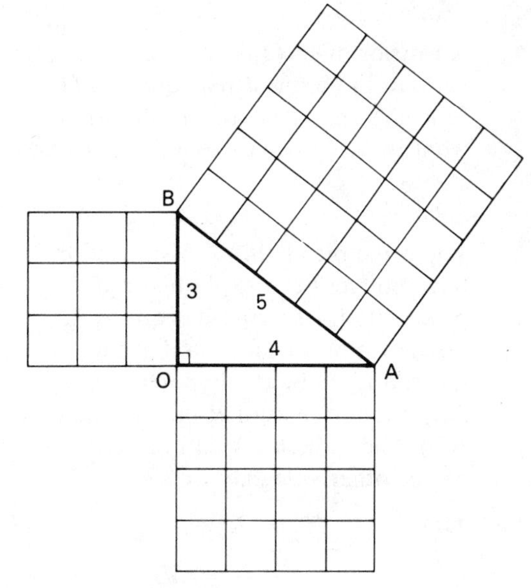

Figure 1.3 Pythagoras' Theorem

To check, $5^2 = 4^2 + 3^2$

i.e. $25 = 16 + 9$

which is correct.

Example 1.4

A ladder 4 m long rests against a wall of a house so that the foot of the ladder is 1 m from the bottom of the wall. How far up the wall does it reach?

Solution

Providing the wall is at right angles to the ground we can use Pythagoras' Theorem. Thus, the ladder length must represent the hypotenuse (say AB) and the base length (say AO) is the distance of the ladder from the foot of the wall. The question asks us to find the perpendicular length (say BO).

If $AB^2 = AO^2 + BO^2$

then $BO^2 = AB^2 - AO^2$

Thus $BO^2 = 4^2 - 1^2$

$= 15$

and $BO = \sqrt{15}$

$= 3.87$ m (to 3 sig. figs.)

Example 1.5

A resistor of 15 Ω is connected in series with an inductor of reactance 20 Ω. Ignoring any resistance in the inductor, what is the impedance of the circuit?

Solution

The principles behind this question have not yet been explained but the procedure for solving it requires no transposition of formula. Let the impedance Z be the hypotenuse (say AB) and the resistor R be the side (say AO). The inductor X can be represented by the other side (say BO).

Thus $\quad Z^2 = R^2 + X^2$

$$= 15^2 + 20^2$$

$$= 225 + 400$$

$$= 625$$

and $\quad Z = \sqrt{625}$

$$= 25\ \Omega$$

A brief mention will now be made of some other types of triangle.

triangle: a figure enclosed by three straight lines

acute-angled triangle: a triangle with three acute angles, i.e. all are less than a right angle

isosceles triangle: a triangle with two sides equal lengths and two equal angles

obtuse-angled triangle: a triangle with one angle greater than a right angle

equilateral triangle: a triangle with three equal sides and three equal angles

Let us consider the equilateral triangle in more detail. Such a triangle is shown in Figure 1.4 and you should construct it yourself, making each side 6 cm long. For accuracy use compasses and check that all the angles are 60°. Then letter the figure as shown. The line marked h is called the perpendicular, since it is at right angles to the base line BC. You can measure the length of this line using a ruler, or calculate it using Pythagoras' Theorem. It is approximately 5.2 cm long. In an equilateral triangle the line h is always 0.866 ($\sqrt{3}/2$) times

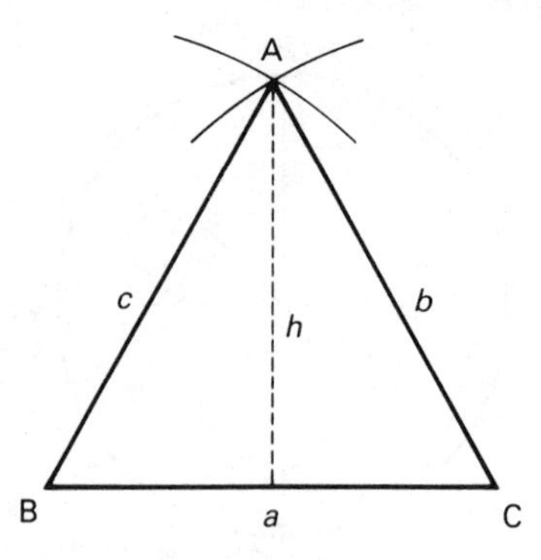

Figure 1.4 Equilateral triangle

shorter than the sides. Now the area of any triangle can be found by multiplying half the base measurement BC by the perpendicular height h. Thus

$$\text{Area } A = \tfrac{1}{2}\,\text{Base} \times \text{Height}$$

$$= \tfrac{1}{2}(6 \times 5.196)$$

$$= 15.59\ \text{cm}^2 \quad \text{(to 2 decimal places)}$$

An alternative method for determining the area of any triangle is to use the formula

$$A = \sqrt{s(s-a)(s-b)(s-c)}$$

where $\quad s = \tfrac{1}{2}(a+b+c)$

For example, in the above question,

$$s = \tfrac{1}{2}(6+6+6) = 9$$

and therefore

$$A = \sqrt{9(9-6)(9-6)(9-6)}$$

$$= 15.59\ \text{cm}^2$$

Other types of plane figures are

quadrilateral: a figure bounded by four straight lines

parallelogram: a quadrilateral with opposite sides parallel and of equal length

rectangle: a parallelogram with all four angles right angles

square: a rectangle with all four sides of equal length

rhombus: a parallelogram with all four sides of equal length, and the angles not right angles

trapezium: a quadrilateral with one pair of opposite sides parallel

Polygons (figures bounded by any number of straight lines) will not be discussed in this book.

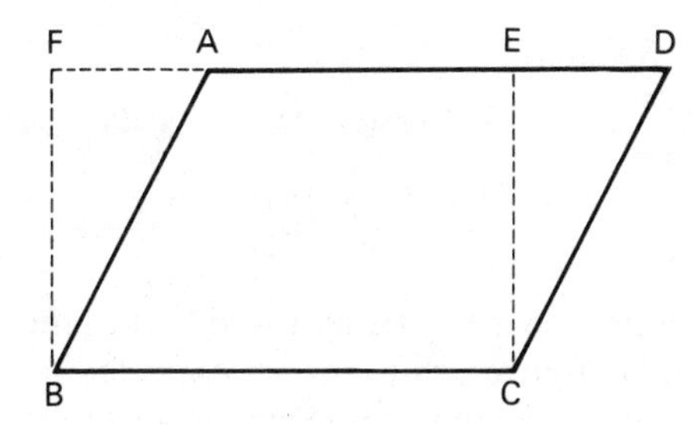

Figure 1.5 Parallelogram

The areas of squares and rectangles can be found easily by multiplying Length × Breadth, but in a parallelogram the figure needs to be transformed into a rectangle to find its perpendicular height. Figure 1.5 shows a typical parallelogram, labelled ABCD. With the perpendiculars BF and CE drawn, the figure can be pushed back to become a rectangle and its area calculated by multiplying its base by the perpendicular height. Thus

$$\text{Area} = \text{AD} \times \text{CE}$$

Note If a diagonal line were drawn between A and C, the parallelogram could be treated as two separate triangles and the area found by

$$\text{Area} = 2 \times \tfrac{1}{2}\text{AD} \times \text{CE}$$

To find the area of trapeziums, try to transform them into rectangles or even triangles and add the separate areas together.

The **circle** shown in Figure 1.2 is by definition a plane figure contained by a line described as the circumference. The ratio of the circumference C to the diameter d is denoted by **pi (π)**. Pi is the circle's constant and is quite commonly taken as 3.142. The circumference of any circle can be found from the formula

$$C = \pi d$$

Since the diameter is twice the radius, then

$$C = 2\pi r$$

You are not expected to know the mathematical proof of the area of a circle. It is sufficient to say that the length of an arc may be measured approximately by dividing it into very short portions, and measuring these lengths as if they were straight. The area of the circle can be found by adding together all the areas of the small isosceles triangles thus formed (i.e. the sum of all the $\tfrac{1}{2}$Base × Height). Hence the area of a circle can be found by the formula

$$\begin{aligned} \text{Area} &= \tfrac{1}{2} \times 2\pi r \times r \\ &= \pi r^2 \end{aligned}$$

Example 1.6

A copper busbar measures 5 cm in diameter. Find the length of its circumference and its cross-sectional area.

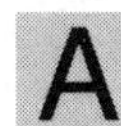

Solution

From the above formula

$$\begin{aligned} C &= \pi d \\ &= 3.142 \times 5 \\ &= 15.71 \text{ cm} \end{aligned}$$

Also

$$\begin{aligned} A &= \pi r^2 \\ &= 3.142 \times 2.5^2 \\ &= 19.63 \text{ cm}^2 \end{aligned}$$

Note It should be pointed out that the area of a circle (πr^2) can be expressed in terms of its diameter (d) instead of its radius (r).

Since $$r = \frac{d}{2}$$

and $$r^2 = \left(\frac{d}{2}\right)^2 = \frac{d^2}{4}$$

then $$A = \frac{\pi d^2}{4}$$

Using this formula is often much quicker than getting involved with radius measurements, as will be seen in the following example.

Example 1.7

A metal conduit has an outside diameter D of 20 mm and an inside diameter d of 15 mm. What is the rim area of the conduit?

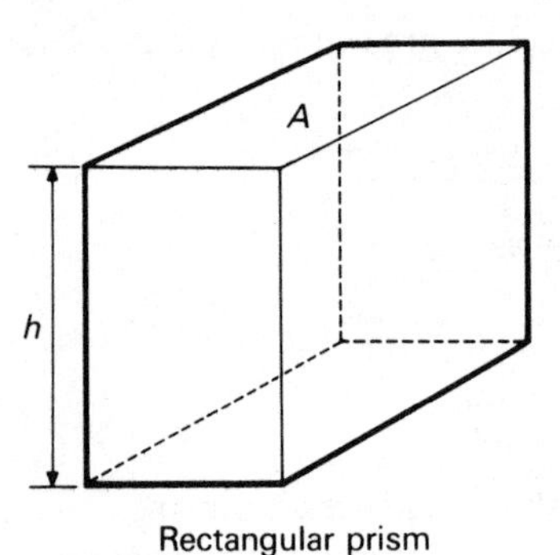

Circular prism (cylinder)

Figure 1.6 Regular solids

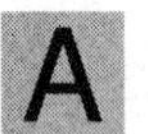

Solution

This example is solved by subtracting the conduit's inside area from its outside area. Thus

$$A = \frac{\pi D^2}{4} - \frac{\pi d^2}{4}$$

$$= \frac{\pi}{4}(D^2 - d^2)$$

$$= 0.785(20^2 - 15^2)$$

$$= 137.4 \text{ mm}^2$$

You should also know how to find the volume of regular solids such as the rectangular prism and cylinder. The rectangular prism is a solid with two rectangular plane ends of the same size and shape, whereas the cylinder is a solid with two parallel circular ends (see Figure 1.6). In both cases, the volume V is the product of the base area A and height h, i.e.

$$V = Ah$$

If the volume of a hollow cylinder is required the following formula is used

$$V = \frac{\pi h}{4}(D^2 - d^2)$$

where D is the diameter of the outside and d is the diameter of the inside. If the total surface area S of a cylinder is required the formula is

$$S = 2\pi r(h + r)$$

Example 1.8

A solid copper busbar is 0.5 m long and 2 cm in diameter. What is its total surface area and volume?

Solution

$$S = 2\pi(50 + 1) = 320.48 \text{ cm}^2$$

$$V = \pi 50 = 157.1 \text{ cm}^3$$

You will not normally be required to find the areas of irregular figures, but you are expected to know how to determine the average value and root mean square value of an a.c. sinusoidal wave. Figure 1.8 shows how an a.c. sine wave is produced. It will be seen that its average value over the complete cycle is zero, but for half a cycle, as would be obtained from a rectified supply, it is found to be the area under the curve, divided by the length of the base. One method of obtaining this value is using the **mid-ordinate rule**, where the lengths of mid-ordinates e_1 to e_n are measured from the base axis to their points of intersection on the curve. By dividing by the number of mid-ordinates n, the average value is found. Theoretically this should be 0.637 × maximum value. For an a.c. voltage, the formula to use is as follows:

$$\text{Average value} = \frac{(e_1 + e_2 + \ldots + e_n)}{n}$$

This value in a.c. theory is of little importance. It is the effective value or root mean square value that is important and which is indicated by measuring instruments. It is expressed in terms of the direct current that produces the same heating effect in a resistor and is found by the formula

$$\text{r.m.s. value} = \sqrt{\frac{(e_1^2 + e_2^2 + \ldots + e_n^2)}{n}}$$

Theoretically this value should be 0.707 × maximum value.

Example 1.9

The maximum value of an a.c. sine wave voltage is 339.46 V and that of an a.c. sine wave current is 20 A. Determine their r.m.s. values.

Solution

$$V_{\text{r.m.s.}} = 0.707 \times 339.46 = 240 \text{ V}$$

$$\text{and } I_{\text{r.m.s.}} = 0.707 \times 20 = 14.17 \text{ A}$$

It will be noticed that the sine wave in Figure 1.8 only covers one cycle. In practice, the public supply provides 50 cycles every second, which is a frequency f of 50 Hertz. The time taken for one cycle is called the periodic time T and equals $1/f$. For 50 Hz, the cycle lasts 0.02 s.

Graphs and charts

A graph is a diagrammatic way of displaying data to compare related quantities. It is generally plotted on axes at right angles to each other. The base **(x-axis)** is called the **abscissa** and the perpendicular **(y-axis)** is called the **ordinate**. It is important to choose suitable scales for both axes and these should be clearly marked and given titles of the quantities they represent. The scales need not start from zero but they should be sufficiently extended so that the graph is not confined to a small part of the graph paper. If a graph is drawn in the first quadrant of Figure 1.2, both the x and y axes have positive values. In the second quadrant the x-axis has a negative value while the y-axis remains positive; in the third quadrant both axes have negative values; in the fourth quadrant the x-axis returns to a positive value while the y-axis remains negative.

Let us now consider some of the common types of graph found in electrical science studies. Probably one of the first graphs you will draw in your first year will be the two axes graph of Ohm's Law, where a linear resistor is used to demonstrate the relationship between potential difference and current. The graph of these two quantities should show a direct variation and be a sloping straight line rising from zero, as shown in Figure 1.7. It tells us that the potential difference across the ends of the resistor is proportional to the steady current passing through it. Graphs which start off with high values and then slope downwards denote inversely proportional characteristics such as are found in the relationship between resistance and cross-sectional area, and between capacitive reactance and frequency. Graphs that are straight horizontal or vertical lines indicate that one of the quantities is unaffected by any variation in the other quantity. Some graphs rise up proportionally and then start to bend over and gradually level out. These often show that a point of saturation has been reached, where the y values are no longer increasing at the same rate as the x values. An example of this is B–H magnetisation curves.

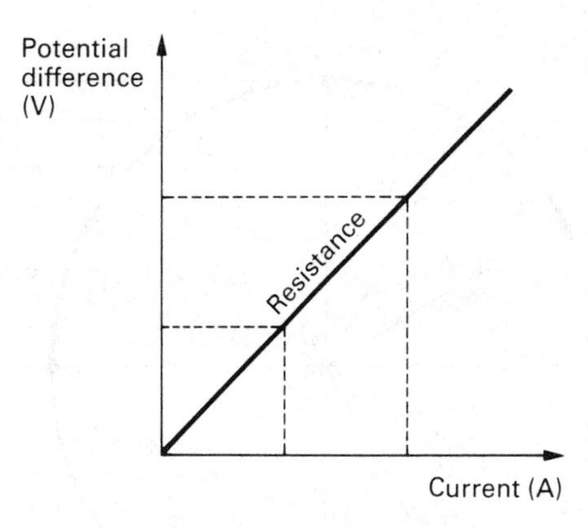

Figure 1.7 Graph showing Ohm's Law

Since reference has already been made to the circle, the graph of a sine wave over one complete cycle of 360° is shown in Figure 1.8 (overleaf), which illustrates how an induced e.m.f. is created by a rotating conductor situated in a uniform magnetic field. The x-axis represents the periodic time of this wave and the scale is marked in twelve intervals, 1 cm apart, each representing 30°. The y-axis represents the induced e.m.f. and is given an overall scale measurement of 6 cm (3 cm in the positive direction and 3 cm in the negative direction) controlled by the circle's radius. The sine wave produced is cyclic in nature, reaching two maximum values, one at 90° and the other at 270°. The thirteen points marked on the graph are called **instantaneous points** and you will notice that the induced e.m.f. passes through zero at 0°, 180° and 360°. Other properties of this a.c. sine wave will be discussed in Chapter 4.

You will not be required to produce circle charts (pie charts) or Venn diagrams but you need to know the procedure for drawing bar charts, which are similar to histograms used in statistics and in which frequency distributions are illustrated by rectangles. Bar charts are used in the construction industry to show work operations occurring in large projects in relation to time taken for their completion. In practice, work activities are listed below each other and their inter-relatedness compared; they are drawn from left to right across graph paper with the horizontal axis normally representing days or weeks. Some work activities cannot start before others and the bar charts show when workers on site are needed or can be taken away to other projects when not needed. Bar charts are dealt with more fully in Part II studies.

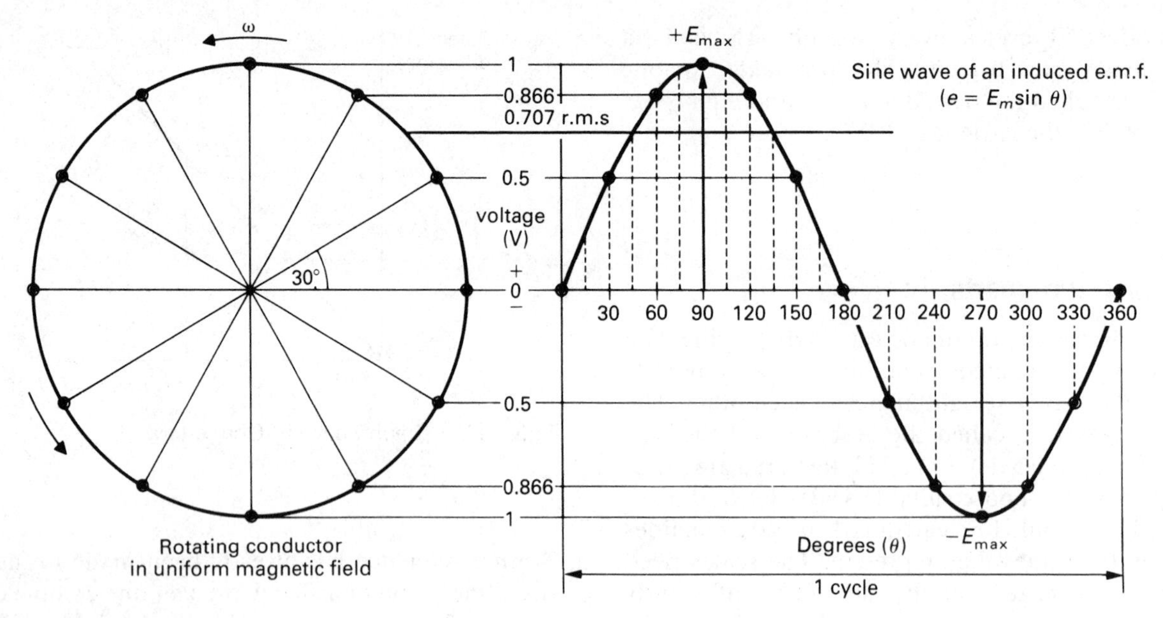

Figure 1.8 Generation of a sine wave

Ratio, proportion and percentage

It is common to speak of a **ratio** as the numerical relation one quantity bears to another of the same kind or units. The sides of a right-angled triangle have already been mentioned and are a good example of ratios, since altering the length of any one side alters the length of the other sides. From an electrical point of view, 100 V divided by 60 V will give a ratio of either 100 : 60, 50 : 30, 10 : 6 or even 5 : 3. Ratios should always be written in the easiest form and it should be pointed out that 5 : 3 is not the same as 3 : 5. A transformer may have a step-down voltage ratio of 5 : 3, but if it was stated 3 : 5 you would expect the transformer to have a step-up voltage ratio.

When two ratios are equal they are called **proportional** ($\propto$). We have already seen that the circumference of a circle is proportional to its diameter. When two circles are compared, we can say that the ratio of circumferences of the first and second circles is equal to the ratio of diameters of the first and second circles, i.e. $C_1/C_2 = d_1/d_2$. The usual way of writing down quantities which are proportional to each other is to insert the proportional symbol between them. For example, the resistance R of a wire is proportional to its length l, but it is inversely proportional to its cross-sectional area A. This is normally written

$$R \propto \frac{l}{A}$$

Table 5 Miscellaneous terms and symbols

Terms	*Symbols*	*Often used to signify*
Greek letters		
alpha	α	angle, or temperature coefficient of resistance
eta	η	efficiency
mu	μ	micro
omega	Ω	ohm, resistance
phi	Φ, ϕ	magnetic flux, angle, or phase difference
pi	π	circle constant – circumference/diameter
rho	ρ	resistivity
sigma	Σ, σ	sum of, or conductivity
theta	θ	angle or temperature
Signs		
	$\approx$	approximately equal to
	$\propto$	proportional to
	∞	infinity
	$>$	greater than
	$<$	less than
	$\geqslant$	equal to or greater than
	$\leqslant$	equal to or less than

A **percentage** is the fractional part of a quantity with 100 as the denominator. Something which is 100% efficient has no losses. In cable problems you are often asked to find the voltage drop in circuits, where the maximum allowance is based on 2.5% of the declared voltage. So if the supply voltage is 240 V, then 6 V is the maximum that can be lost at load terminals. This can be written as

$$\text{Voltage drop } (V) = \frac{2.5}{100} \times 240 = 6 \text{ V}$$

Often manufacturers of machines and switchgear express the operation of their equipment in values above 100%, e.g. 150%, 200%, 300%, etc. This provides information about the apparatus, especially its overcurrent allowance. Since 100% = 1, the percentages above should be interpreted as 1.5, 2 and 3 times the normal condition. It is common practice to interpret results in per unit values rather than in a percentage form. For example, a motor's rotor might have a percentage slip of 5% which is written as 0.05 per unit.

EXERCISE 1

1. Convert the following:

a) 0.35 ohms into milliohms
b) 750 millijoules into joules
c) 255 kilowatts into megawatts
d) 400 000 volts into kilovolts
e) 0.022 millifarads into microfarads.

BS3939 circuit symbols

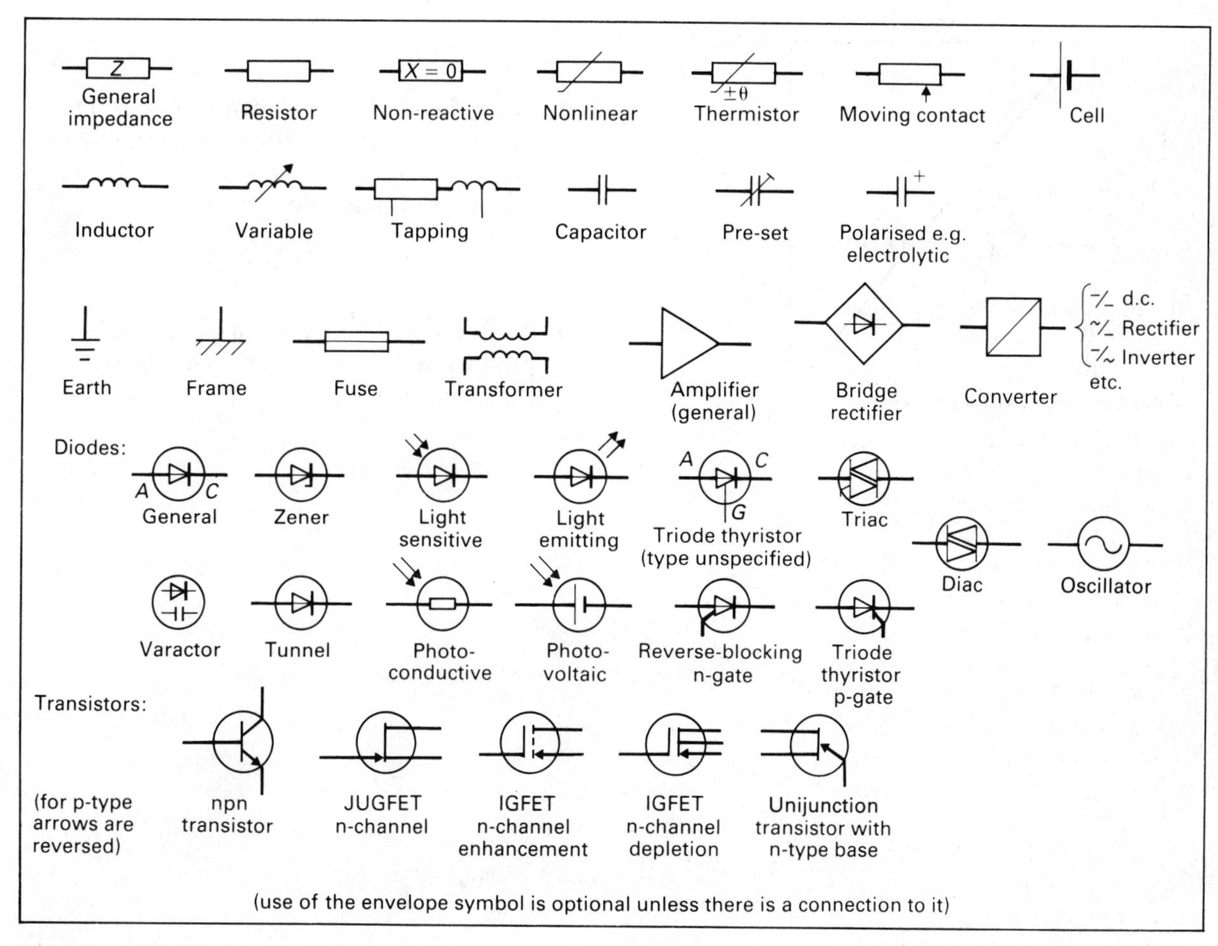

Figure 1.9 BS3939 circuit symbols

2. Write down the signs or Greek letters for the following terms and briefly explain their use:

a) efficiency
b) pi
c) resistivity
d) proportional to
e) equal to or greater than.

3. a) Transpose the formula $A = \frac{\pi d^2}{4}$ to find d.

b) Calculate the diameter of a 2.5 mm^2 cable.

4. Solve a) $10^6 \times 10^3 \times 10^2$
b) $10^{-5} \times 10^0 \times 10^3$

5. a) Using Pythagoras' Theorem, determine the value of V_s in Figure 1.10.

b) Using a protractor, determine the angle ϕ in Figure 1.10.

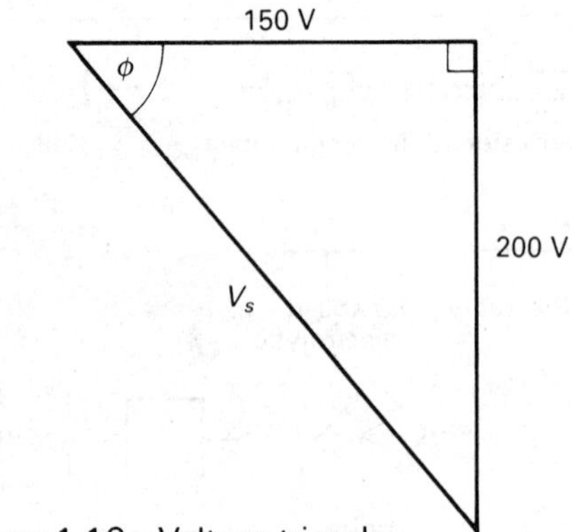

Figure 1.10 Voltage triangle

6. a) Determine the area of a triangle that has sides 3 cm, 4 cm and 5 cm.

b) How many degrees are there in a right-angled triangle?

7. What is the volume of a copper busbar of length 3 m, having inside diameter 2.4 cm and outside diameter 3 cm?

8. The results table below shows the voltage and current readings obtained in an experiment to find the value of an unknown resistor. Plot the results on graph paper and determine the resistor's value from the slope of the line.

V	4.4	10	14	18.8	22
I	2.0	4	6	8	9

9. Draw the sine wave shown in Figure 1.8 over half a cycle and from it determine its average and root mean square values, using the mid-ordinate rule.

10. A 20 : 1 step-down transformer has a primary voltage of 6.6 kV. What is its secondary voltage?

Mechanical science 2

Objectives

After reading this chapter you should be able to:

- *describe and state the differences and relationships between mass, force and weight.*
- *complete exercises on mass, force and weight.*
- *describe force as a vector quantity and describe the conditions for equilibrium.*
- *describe the principle of moments and carry out simple calculations.*
- *define density as mass per unit volume and carry out calculations.*
- *distinguish between heat and temperature and carry out simple calculations involving the coefficient of linear expansion.*
- *define work as Force × Distance and apply this in simple calculations.*

Mass, force and weight

Mass (m) is the amount of matter contained in a body. Its unit of measurement is the kilogram (kg) and it remains constant once it has been fixed. A body's mass can easily be found by comparing it on balance scales with standard known masses marked in either grams or kilograms.

Force (F) is an external agency capable of altering the state of rest in a body through a 'push' or 'pull' action. Its unit of measurement is the **newton** N, named after Sir Isaac Newton, defined as the force required to give a mass of one kilogram an acceleration of one metre per second per second (m/s^2). Force may manifest itself in many different ways.

inertial force: the force needed to stop and start things, to change direction, or velocity

cohesive/adhesive forces: forces that hold things together

electrical force: force created by an electric field

magnetic force: force created by a magnetic field

tensile force: force that places a material under tension, tending to stretch it. For example, the force on the wire ropes on a loaded crane.

compressive force: force that squeezes or compresses a material, e.g. the force on the spindle of a screw jack lifting a load

shearing force: force that moves one face of a material over an adjacent face, e.g. the force on a rivet holding metal plates together

friction force: force capable of resisting or even preventing the relative movement of two surfaces that are in contact with each other

centripetal force: force acting towards the centre when a mass attached to a string is swung round in a circular path

centrifugal force: the opposite force to the centripetal force in the string (above)

gravitational force: the force due to gravity

Figure 2.1 illustrates some examples of different forces.

Weight (W) is the force produced by gravity acting on a mass, and its unit of measurement is the newton. If a mass of 1 kg is suspended on a string, there is a force of gravity on the mass acting downwards and a tensile force in the string acting upwards, to hold the mass in equilibrium. If the string is broken the mass accelerates towards the centre of the Earth with an acceleration g of 9.81 m/s^2. If the 1 kg mass is taken into outer space it becomes 'weightless' but back on Earth under the influence of gravity its weight is 9.81 N. Force due to gravity is given by the formula:

$$F = mg \qquad [2.1]$$

where g is the acceleration of free fall, 9.81 m/s^2.

Note that the above is an engineering/scientific definition of weight. If you go into a shop and ask

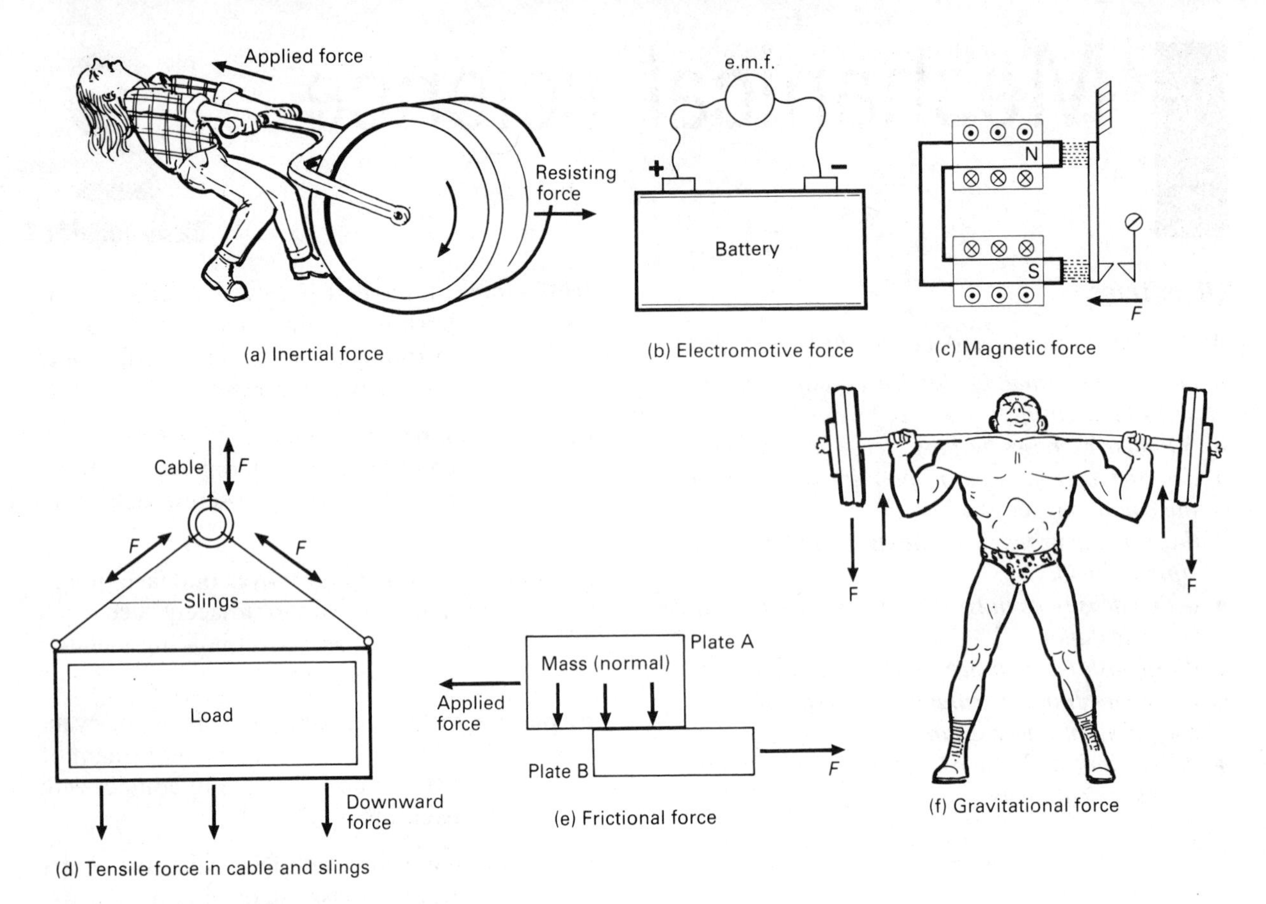

Figure 2.1 Types of force

for 50 N of apples the shopkeeper will not understand what you want. The lay person confuses weight and mass and refers to both in kilograms (the apples actually weigh 50 N which is approximately 5 kg).

Newton's laws of motion

Newton stated three laws of motion which can be interpreted as follows.

1. Every body remains at rest or moves with constant velocity in a straight line unless an external force acts upon it.
   This means that a body will not move until it is pushed.
2. The acceleration which results in a change of velocity of a body is directly proportional to the force applied to the body and takes place along the straight line in which the force acts.
   This means that the harder you push a body, the greater its acceleration.
3. To every applied force there is an equal and opposite reacting force.
   This means, push me and I'll push you back!

In Newton's first law, you have to consider that all bodies are inert and require a force to cause them to move. The reluctance to start moving from rest, or stop moving when an opposing force is applied, is called **inertia** (as previously mentioned). Once a body is moving, it will tend to keep on moving in the same direction. This is illustrated simply by the movement of passengers in a car. When the brakes are applied sharply the passengers keep moving in the direction in which the car is travelling. If the front seat passengers are strapped in then the opposing force is applied by the seat belts. If the back seat passengers are *not* strapped in, then they may overtake the driver and go through the windscreen. Also worth noting is that a heavy body requires more effort to make it change from one speed to another than a lighter body, irrespective of their sizes.

In Newton's second law, the words **acceleration** and **velocity** are used. These two terms are **vector quantities**, i.e. they have magnitude and direction (see 'Vector quantities' on p.18). Acceleration a is defined as the rate of change of velocity, given by the expression

$$a = \frac{\text{Change in velocity (m/s)}}{\text{Time taken for change (s)}} \qquad [2.2]$$

Velocity v is almost the same thing as speed. It has both magnitude and direction, whereas speed does not have direction. It is the rate of change of position, i.e. distance travelled divided by time taken. It can be shown that since the rate of change in velocity is proportional to the applied force and to the acceleration of a body, then force is proportional to acceleration, i.e.

$$F \propto a \qquad [2.3]$$

Since the mass of a body is the property of the body to resist change of motion then force is also proportional to mass, i.e.

$$F \propto m \qquad [2.4]$$

Combining these two expressions gives

$$F \propto ma \qquad [2.5]$$

Newton discovered that forces always occur in pairs. Whenever a force acts on a body, there must be another force, equal in size but opposite in direction, acting on another body. This is the basis of his third law. These forces are called **action** and **reaction** forces. They can be demonstrated by the simple example of resting your elbow on a table. The downward force of your elbow is equal to the upward force of the table (see Figure 2.2). Further examples of this law are given in Exercise 2.1.

Force causes strain and this may place a body or material in a state of stress, which is the ratio of force applied to the area over which it acts.

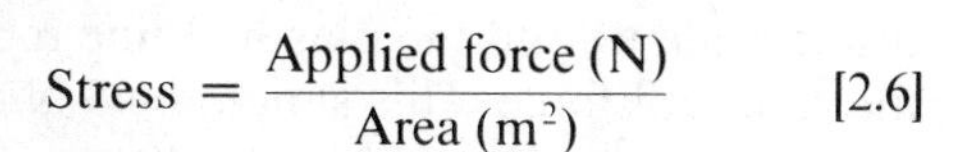

$$\text{Stress} = \frac{\text{Applied force (N)}}{\text{Area (m}^2\text{)}} \qquad [2.6]$$

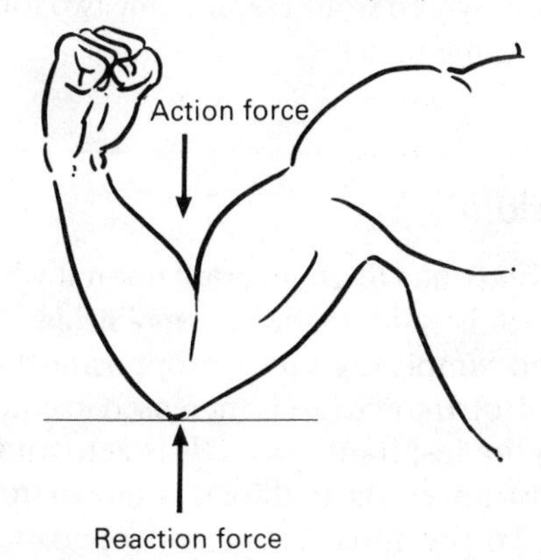

Figure 2.2 Opposite forces

Within a material's **elasticity** or stretching limits, stress is found to be directly proportional to strain. This can be illustrated simply by stretching a helical spring, where the extension is found to be directly proportional to the load hanging on it.

EXERCISE 2.1

1.

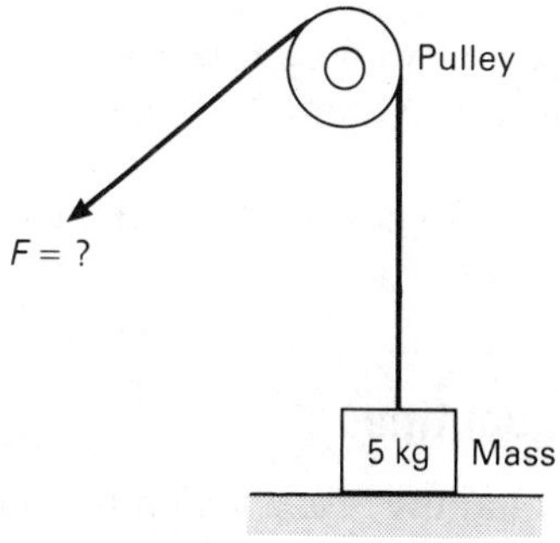

Figure 2.3 Lifting a load

In Figure 2.3, what is the force required to lift the load?

2. Make labelled sketches to illustrate the following forces:

a) gravitational force
b) magnetic force
c) electrical force
d) inertial force.

3. In an experiment with a spring the following results were obtained:

Load (N)	50	100	150	200
Spring stretch (cm)	9	11	13	15

Draw a graph of these results and find

a) the length of the spring when it is not stretched;
b) the length of the spring for a load of 80 N;
c) the load needed to produce an extension of 12 cm.

4. What force must be applied to a mass of 2 kg to accelerate it at 12 m/s^2?

5. Determine the force applied to a mass of 50 mg if it is accelerating at 5000 m/s^2.

6. A truck of mass 400 kg is pushed along a horizontal track with a force of 40 N. Ignoring friction, determine the truck's acceleration.

7. Work (which is energy) is an expression of force times distance. In Figure 2.3, what work is required to lift the load 2 m from the ground?

8. Calculate the stress in a 2.5 mm^2 bare copper cable if a tensile force of 15 N is applied.

9. Explain in your own words the difference between mass and weight.

10. Explain friction force, giving *four* practical examples.

Vector quantities

A vector quantity is one which has **magnitude** and **direction**, such as force. If you want to move an object from one place to another you have to apply a force in the required direction. Vectors are lines drawn to represent quantities, such as velocity, force, etc. The length of the line represents the size or magnitude of the quantity and the direction of the line indicates the direction in which the quantity acts. If two forces act at a point in the same direction along the same line of action, the resultant force is the sum of the two forces. If they act at a point in opposite directions along the same line of action, the resultant force is the difference between the two forces (see Figure 2.4).

If two forces act in different directions at the same point, like two ropes pulling on a post, their

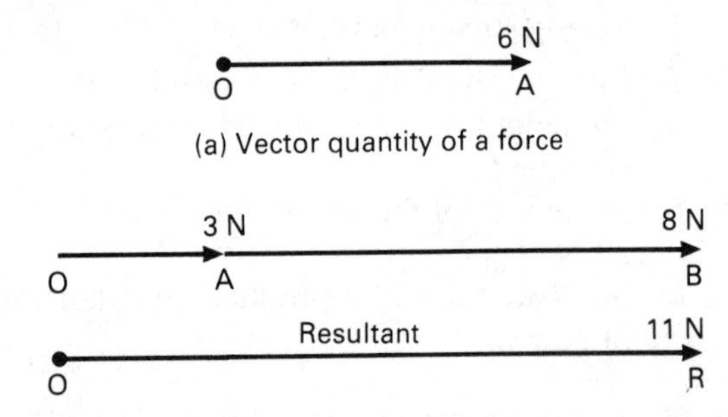

(a) Vector quantity of a force

(b) Two forces acting in the same direction are **added**

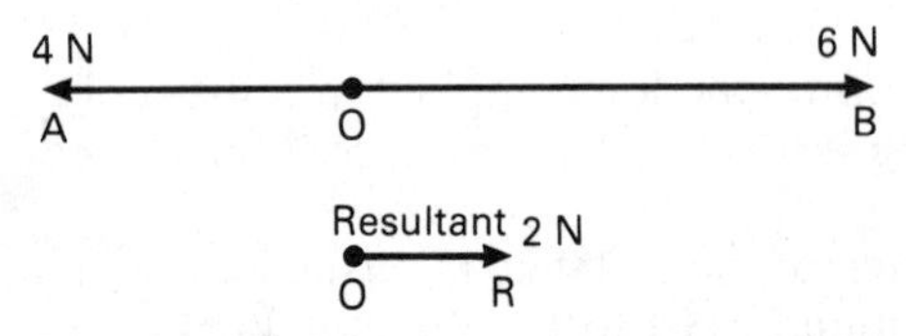

(c) Two forces acting in opposite directions are **subtracted**

Figure 2.4 Vector quantities

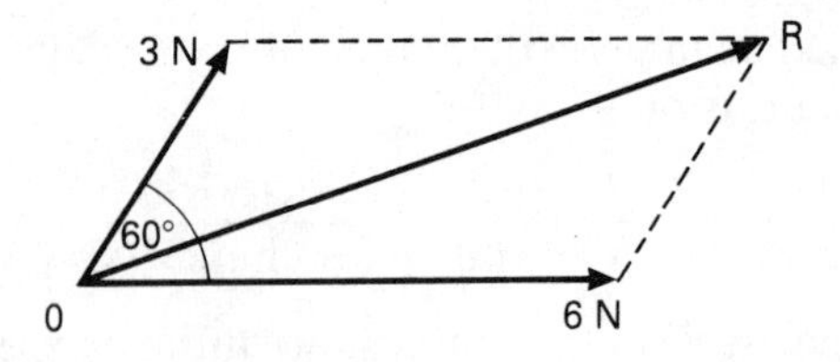

Figure 2.5 Parallelogram of forces

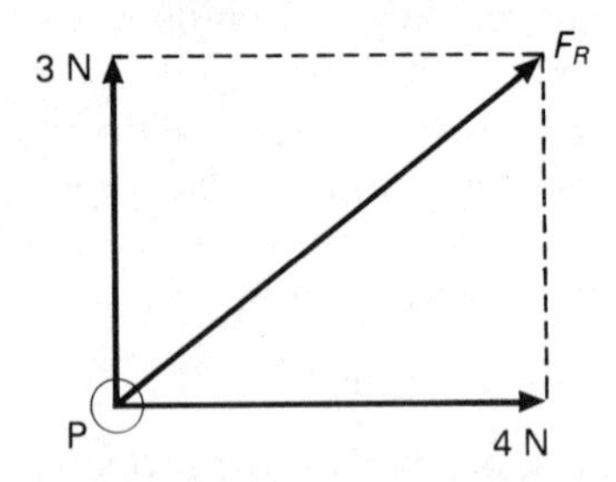

Figure 2.6 Resultant force of 3 N and 4 N

resultant can be found by the method known as **parallelogram of forces**. Figure 2.5 shows two forces of 3 N and 6 N acting at 60° to each other. To construct this diagram, first choose a suitable scale, say 1 cm = 1 N, and then draw both forces at the correct angle, using arrows to show direction. Complete the parallelogram with dotted lines and then draw in the diagonal. The diagonal shows the direction and magnitude of the resultant force. If you measure the resultant OR against your chosen scale, you should find the magnitude of the resultant force to be 7.9 N.

Consider the example in Figure 2.6, which shows an aerial view of two overhead electricity cables attached to a post P, with horizontal pulls of 4 N and 3 N respectively. If the angle made between the cables is 90°, what is the resultant pull on the post? Copy this diagram using an appropriate scale to find the diagonal line representing the resultant force. This should be 5 N and is the single force which would replace the two forces and have the same effect.

Equilibrium

When a body has forces exerted upon it which just balance, it is said to be in **equilibrium**. This is illustrated simply by the two opposing forces in Figure 2.4(c). If force OA is increased to equal force OB then the resultant force OR is zero and a state of equilibrium exists. In the cases shown in Figures 2.5 and 2.6 the force equal and opposite to the resultant is called the **equilibriant** and is the single force which would be needed to hold the two applied forces in equilibrium.

If a stay wire is attached to the post in Figure 2.7, making an angle of 45° with the post, it will be seen that a tension force of approximately 7 N is needed in it to balance the pull of the electric cable.

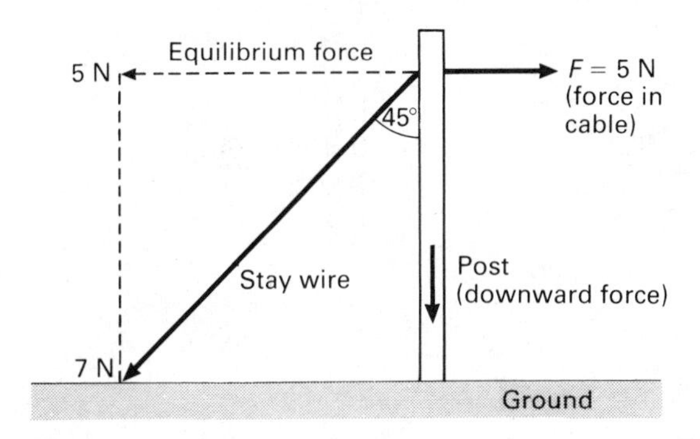

Figure 2.7 Tension in stay wire

When three forces act at a point on a body to hold it in equilibrium, they can easily be represented by vectors forming the sides of a triangle. This method is known as the **triangle of forces** and it requires the construction of two diagrams, a **space diagram** that provides the physical details of the problem and a **force diagram**.

Consider a load of 100 N supported by two ropes 4 m and 3 m long, attached to two brackets 5 m apart in a horizontal line. What is the tension in each rope? Figure 2.8 shows how the space and force diagrams should be drawn. The space diagram is marked using capital letters (ABC) in a clockwise order. The force diagram is then constructed as a triangle with lines parallel to the forces in the space diagram. The letters A and B appear on either side of the 100 N force acting downwards. This line is drawn first, to a suitable scale, say 1 cm = 20 N and is marked in lower case letters *ab*. From *b* a line is drawn parallel to BC and labelled *bc*. Its length is not yet known. From *a* a line is drawn parallel with AC, cutting across *bc*. This line is marked *ac*. Each rope tension in newtons is found by measuring *bc* and *ac* and multiplying by 20.

The expression **centre of gravity** is associated with equilibrium. It can be described as the point through which the whole weight of an object seems to act. A suspended plumbline is a good example of this. If it is pushed to one side its centre of gravity rises and gravity tries to pull it back to to its lowest position again. Objects can be stable or unstable, or they may even be neutral like a round object on a

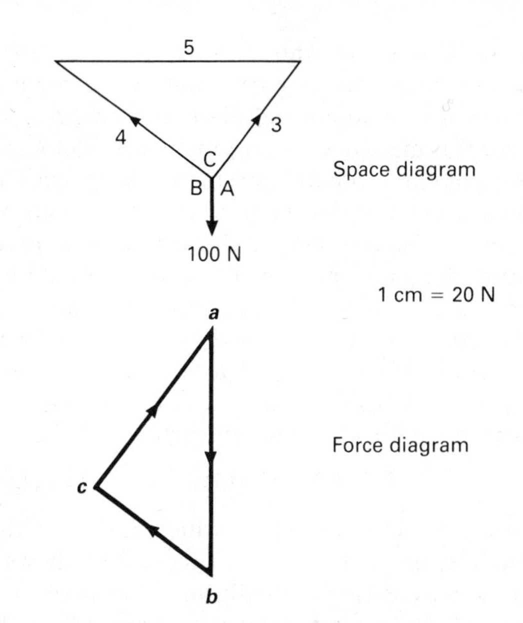

Figure 2.8 Triangle of forces

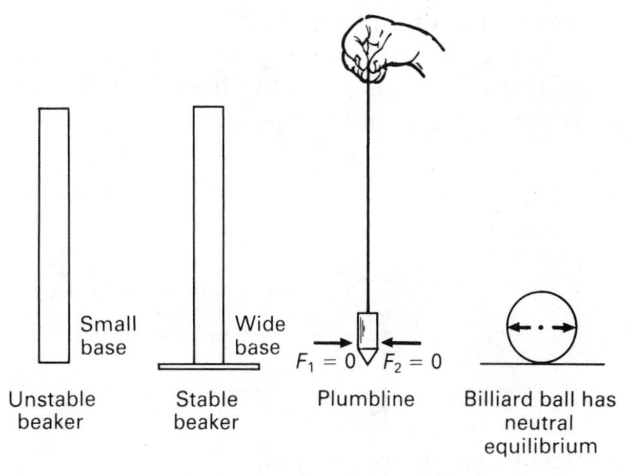

Figure 2.9 Stability of objects

flat surface. Knowing how or where to lift something, or even erect or stack something, relies heavily on knowing where the centre of gravity is. It is important to remember that for an object to be more stable it should have a low centre of gravity and preferably a wide base. A tall object with a small base will quite easily topple over if a small force is applied to one side (see Figure 2.9).

Principle of moments

The turning effect of a force is called a **moment** and is found by the product of **force** and **perpendicular distance** from the force to the pivot or fulcrum, i.e.

$$\text{Moment} = \text{Force (N)} \times \text{Distance (m)}$$

Figure 2.10 shows the turning moment on a spanner about the axis of a nut. Since force is expressed in newtons and distance in metres, the moment is measured in newton metres. A moment can be applied without movement taking place, in which case no work is done, since work is defined as Force × Distance moved. However, an important feature of a moment is the direction in which the force acts (clockwise or anticlockwise). For equilibrium to be achieved a total clockwise moment (CM) must equal a total anticlockwise moment (ACM). This is called the **principle of moments** and is illustrated in Figure 2.11. Thus

$$\text{CM} = \text{ACM} \qquad [2.7]$$

There are many examples which illustrate this principle, particularly tools. Figure 2.12 shows a tap wrench used for internally threading holes. The moments of the forces exerted by the hands on the tap wrench overcome the moments of the forces of the cutting edges of the tap.

The method of finding an unknown force exerted on a body when the lines of action are known is shown in Example 2.1.

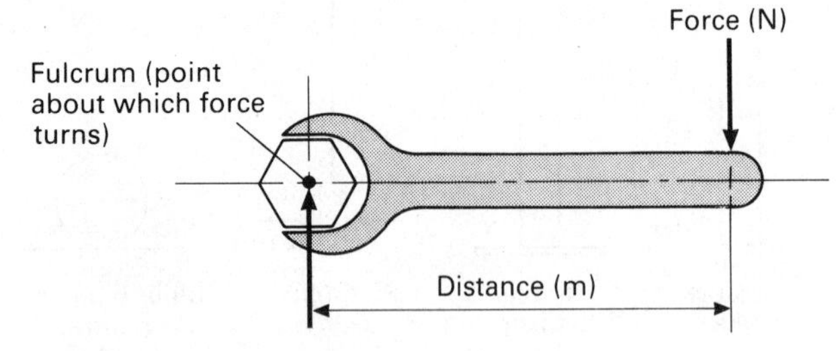

Figure 2.10 Turning moment

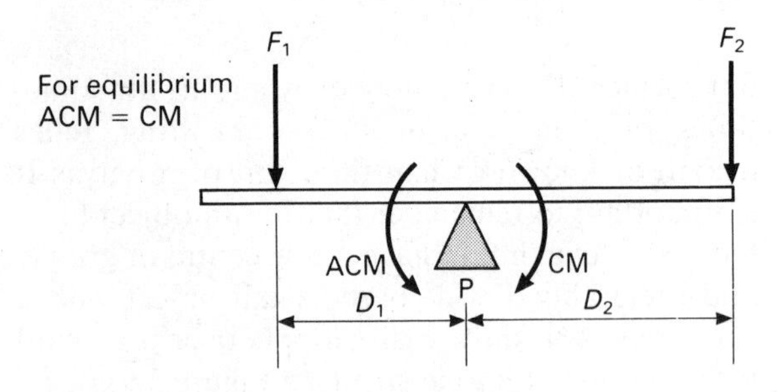

Figure 2.11 Principle of moments

Example 2.1

Figure 2.13 shows two forces acting in opposite directions on the pivot P of a bell crank lever. Calculate the force F_1 to satisfy the conditions for equilibrium.

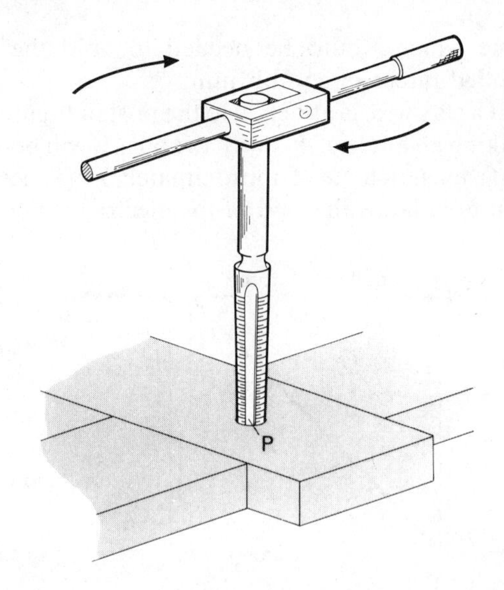

Figure 2.12 Tap wrench demonstrating principle of moments

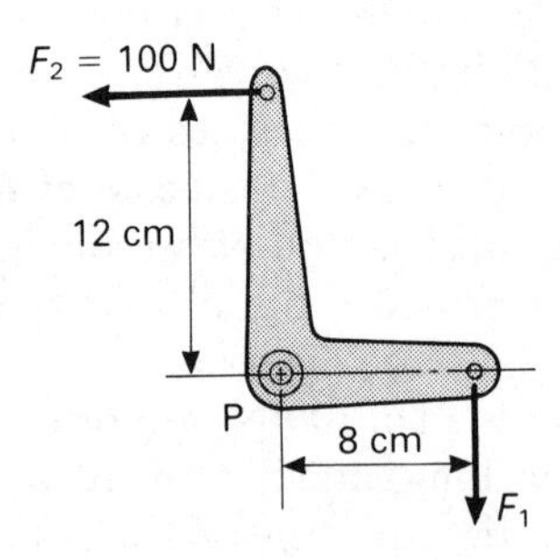

Figure 2.13 Bell crank lever

Solution

From [2.7] CM = ACM

Then

$$\text{Force } (F_1) \times \text{Distance } (D_1) = \text{Force } (F_2) \times \text{Distance } (D_2)$$

$$F_1 \times 8 = 100 \times 12$$

$$F_1 = \frac{1200}{8}$$

$$= 150 \text{ N}$$

Levers also work on the principle of moments. Those where the fulcrum is situated between the effort and load are called Class 1 levers (e.g. a crowbar); those where the fulcrum is at one end and the load is between the fulcrum and effort are called Class 2 levers (e.g. a wheelbarrow); those where the fulcrum is at one end and the effort is

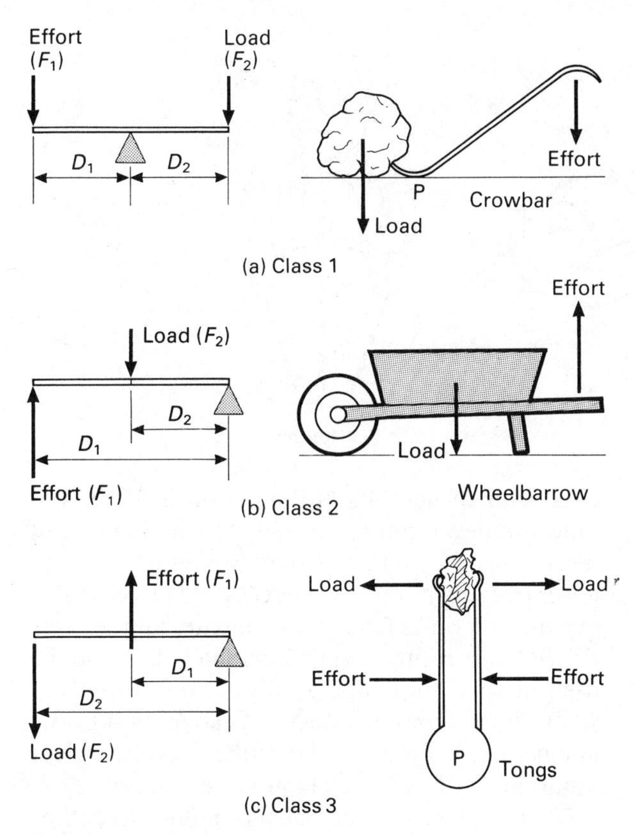

Figure 2.14 Classes of lever

between the fulcrum and load are called Class 3 levers (e.g. coal-tongs). Figure 2.14 illustrates these three classes of lever.

Density

Density (ρ) is the mass (m) per unit volume (V) of a body or substance, measured in kilograms per cubic metre. Thus:

$$\text{Density} = \frac{\text{Mass (kg)}}{\text{Volume (m}^3\text{)}} \quad [2.8]$$

For a given volume the mass of any object will have a specific value. If you compare two rectangular blocks of a different substance, say gold and iron, both of the same dimensions, then although they have the same volume the gold block has a much greater mass, simply because it is denser. Gold has a density of 19 300 kg/m^3, iron of 8 300 kg/m^3, water of 1000 kg/m^3, ice of 920 kg/m^3 and air of 1.3 kg/m^3.

The density of a substance varies with temperature. Often comparisons are made between more dense and less dense substances, and in this case the **relative density** (formerly called specific gravity) is used. This is the ratio of the density of a substance to that of water at 4 °C. At this temperature the density of water is 1000 kg/m^3. Thus, the relative density is a useful way of seeing whether a substance will float or sink in water. For any substance it is given by the ratio

$$\text{Relative density} = \frac{\text{Density of substance}}{\text{Density of water at 4 °C}} \quad [2.9]$$

The relative density of gold is 19.3 and of iron is 8.3. The relative density of ice is 0.92 (less than unity), showing that ice will float on water. Note that these values are stated in numbers and do not have units, since the numerator and denominator are both expressed in kg/m^3.

Relative density can also be found by the ratio of mass of a substance to the mass of the same volume of water.

EXERCISE 2.2

1. What is the density of a lead block of mass 44 000 kg if it has a volume of 4 m^3?

2. An empty room measures 3 m by 4 m by 5 m. If the density of air is 1.3 kg/m^3, calculate the mass of air in the room.

3. A rectangular water tank has sides of 4 m and 5 m. If it can hold a maximum of 40 000 litres of water, what is the height of the tank?

   Note: 1 litre of water has a mass of 1 kg.

4. An electric water heater with a mass of 30 g is placed in a water tank. If the level of water rises from the 50 cm^3 level to 60 cm^3 level what is the density of the heater? What is the mass of the heater in kilograms?

5. What mass of air is contained in a room measuring 2.5 m by 4 m by 20 m if the density of air is 1.3 kg/m^3?

Heat, temperature and expansion

Heat is a form of energy and like other forms of energy it is measured in joules. To give some meaning to this unit, a burning match is likely to produce about 2000 J (2 kJ) whereas a 1 kilowatt electric fire element kept on for one hour (a unit of electricity) is equivalent to 3 600 000 J (3.6 MJ). A

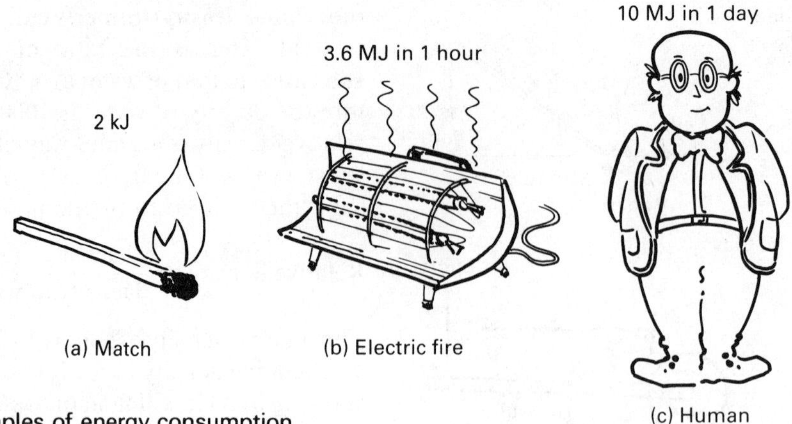

Figure 2.15 Examples of energy consumption

person's body needs about 10 MJ each day to keep going and petrol contains 47 MJ of energy in each kilogram.

Temperature is a state of hotness or coldness of a substance and is measured in degrees Celsius or degrees Fahrenheit. It is a measure of the heat contained in a body, in the sense that to raise the temperature of a piece of steel of mass 10 kg by 20 °C would require twice as much heat as that needed to raise the temperature of a similar piece of steel of mass 5 kg by the same amount.

Figure 2.16 shows a typical mercury-filled thermometer. Mercury is used because it expands evenly as the temperature rises and it is a good conductor of heat. Its boiling point is 357°C but unfortunately it freezes at −39 °C. Another type of thermometer contains alcohol, which has a much lower freezing point that mercury, −112 °C. It also expands six times faster than mercury for the same rise in temperature, but unfortunately it cannot be used in very hot temperatures because it boils at 78 °C. Furthermore, it does not expand uniformly and it sticks to the sides of the tube ; it is also a clear liquid and it has to be coloured to be seen.

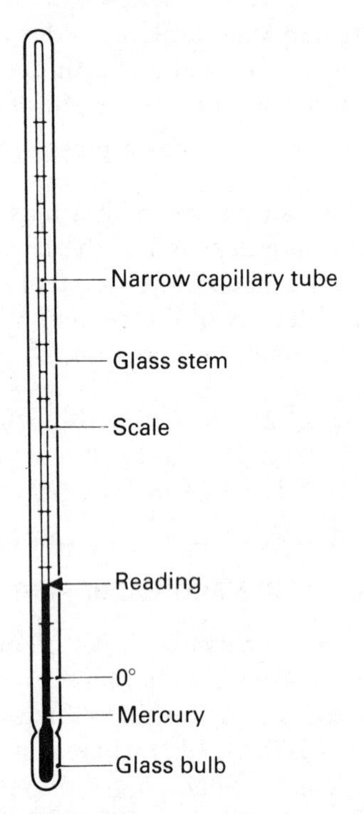

Figure 2.16 Thermometer

There are other types of thermometer such as **resistance thermometers, thermoelectric transducers** (e.g. **thermocouple, thermistor,** etc.) and **gas thermometers**. Many of these are designed to operate over a much wider range of temperature than liquid thermometers.

It is well understood that heat travels by conduction, convection and radiation and when solids, liquids and gases are heated they expand and increase in volume. When they cool down their volumes decrease as long as they do not change from one state to another. For example as water cools its volume reduces, until it changes to ice when it expands. This is why pipes burst when the water freezes and why ice floats on water, since the density of ice is less than that of water. Interestingly, if the density of ice were greater than that of water, it would sink to the bottom of the glass, no convection currents would be formed and our drinks would not get cold. Substances expand because their molecules are vibrating or moving faster and they take up more space. Gases expand more than liquids and liquids expand more than solids. The amount of expansion per unit volume for unit temperature rise is called the **coefficient of cubical expansion**. This varies in different solids and different liquids but remains constant for all gases. For solids the expansion on unit length for

Table 6 Metal characteristics

Melting point of some common metals		*Tempering colour indicators of steel*		*Temperature–colour of hot metals*	
Metal	*Temperature (°C)*	*Colour*	*Temperature (°C)*	*Colour*	*Temperature (°C)*
Aluminium	659	Very pale yellow	200	Dull red	700
Brass	930–1010	Straw yellow	240	Cherry red	900
Bronze	910	Brown yellow	260	Orange	1550
Copper	1083	Light purple	276	Dazzling white	1550
Gold	1075	Dark purple	290		
Lead	327	Dark blue	295		
Mercury	357 (boils)	Pale blue	320		
Silver	961				
Steel	1400–1500				
Tin	232				
Tungsten	3400				
Zinc	419				
Solder (60% tin, 40% lead)	190				

unit temperature rise is called the **coefficient of linear expansion** (α). It is expressed as

$$\alpha = \frac{\text{Change of length}}{\text{Original length} \times \text{Temperature rise}} \qquad [2.10]$$

Brass has a value of 18.7×10^{-6}/°C, while mild steel has a value of 11×10^{-6}/°C. For mercury it is 180×10^{-6}/°C and for oil it is 230×10^{-6}/°C.

Consider the following examples.

Example 2.2

A copper busbar 100 cm long increases in length by 0.17 cm when it is heated through 100 °C by an electric current. What is the copper's coefficient of linear expansion? (See Figure 2.17.)

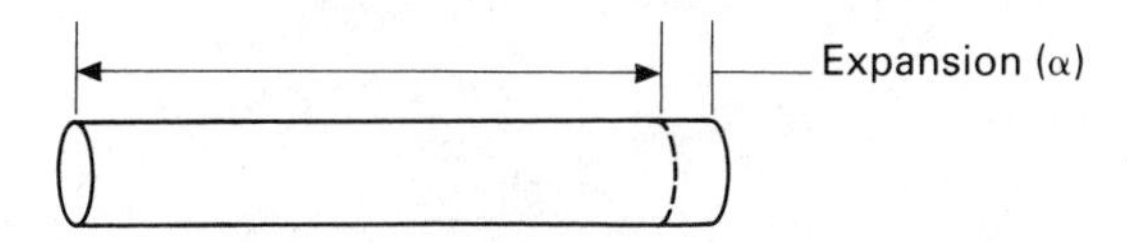

Figure 2.17 Copper busbar

Solution

Using formula [2.10]

$$\alpha = \frac{0.17}{100 \times 100}$$

$$= 0.000\,017 \text{ /°C or } 17 \times 10^{-6} \text{ /°C}$$

Example 2.3

a) Convert 30 °C to °F and 32 °F to °C using the formulae

$$C = \tfrac{5}{9}(F - 32) \quad \text{and} \quad F = \tfrac{9}{5}C + 32$$

b) Convert 411 K into °C and 30 °C into K.

Note: See Chapter 1 for the definition of thermodynamic temperature (K).

c) A bar of aluminium is normally 12 cm long at 20 °C. At what temperature will its length be increased by 1/100 cm if its coefficient of linear expansion is 23×10^{-6} /°C?

Solution

a) $$F = \tfrac{9}{5} \times 30 + 32 = 86 \text{ °F}$$

$$C = \tfrac{5}{9} \times (32 - 32) = 0 \text{ °C}$$

Note: An alternative method is to use the formula:

add 40, times $\tfrac{5}{9}$ (or $\tfrac{9}{5}$), subtract 40

For example,

$$(30 + 40) \times \tfrac{9}{5} - 40 = 126 - 40$$

$$= 86 \text{ °F}$$

Also $(32 + 40) \times \tfrac{5}{9} - 40 = 0$ °C

b) Since 273 K = 0 °C,
then 411 K = 138 °C.
Also 0 °C = 273 K,
so 30 °C = 303 K.

c) Since

$$\text{Expansion} = \text{Length} \times \alpha \times \text{Temp. rise}$$

and

$$\text{Temp. rise} = \frac{\text{Expansion}}{\text{Length} \times \alpha}$$

then

$$t = \frac{0.01}{12 \times 0.000\,023}$$

$$= 32.3\ °\text{C}$$

Therefore the temperature at which the length is increased by 1/100 cm is $20 + 32.2 = 52.2$ °C.

Alloys called **invars** have very low coefficients of expansion and are often used in conjunction with temperature-sensitive bimetals. The oven thermostat shown in Figure 2.18 uses this principle. Briefly, it involves two different metals welded together. The metals have different expansion rates and when they are heated by electric current, one expands more than the other, so the assembly bends and trips out the circuit which it controls. There are many different types of temperature sensing device on the market today, capable of sensing not only temperature but humidity, pressure, force and other quantities. A wide range of these devices are now digital and give a direct reading, rather than operating on an analogue principle, such as a deflecting needle over a marked scale.

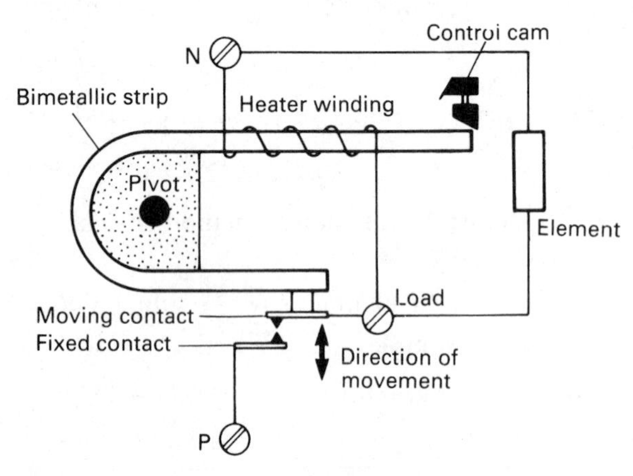

Figure 2.18 Simmerstat

A term associated with heat is called **specific heat capacity** c. The specific heat capacity of a substance is the amount of heat in joules needed to raise the temperature of 1 kg of the substance by 1 °C and it is measured in J/kg °C. Water needs 4200 J to become 1 °C hotter, paraffin 2200 J and ice 2100 J. Water is a good liquid for storing heat, which is the main reason it is used in central heating systems. The quantity of heat Q needed to cause a given temperature rise is expressed as

$$Q = c \times m \times (t_2 - t_1) \qquad [2.11]$$

where m is the mass of substance in kg and t_1 and t_2 are the respective initial and final temperatures, in °C.

Example 2.4

An electric heater is fitted to a copper kettle of mass 0.5 kg. If it is filled with 1.5 litres of water, what is the amount of heat needed to raise the water temperature from 20 °C to boiling point (100 °C)?

Note: For water $c = 4200$ J/kg °C, for copper $c = 380$ J/kg °C, 1 litre of water has mass 1 kg.

Solution

The heat required for the copper kettle is

$$Q_1 = cm(t_2 - t_1)$$
$$= 380 \times 0.5 \times 80$$
$$= 15\,200 \text{ J}$$

The heat required for the water is

$$Q_2 = cm(t_2 - t_1)$$
$$= 4200 \times 1.5 \times 80$$
$$= 504\,000 \text{ J}$$

Total heat required Q is

$$15.2 \text{ kJ} + 504 \text{ kJ} = 519.2 \text{ kJ}$$

Gas laws

In order to understand the gas laws you must first understand the nature of a gas, which is very different from that of a solid or liquid. In a solid, molecules are practically fixed in position, in a liquid they are continuously moving over each other, but in a gas they are moving about at considerable speed, colliding with each other as they are heated. When dealing with gases, three varying quantities have to be considered, namely, **pressure** p, **volume** V and **temperature** T.

Volume and temperature have already been mentioned. Pressure is the force per unit area acting on a surface, measured in newtons per square metre (N/m^2). Pressure in gases may be expressed as the height of liquid that the gas will support. Atmospheric pressure will support 0.76 m (or 760 mm) of mercury. This is known as **standard atmospheric pressure** and is approximately 100 kN/m^2. Other units of pressure that you may come across are: 1 **bar** = 10^5 N/m^2 and 1 **Pascal (Pa)** = 1 N/m^2. Note that atmospheric pressure is approximately 1 bar, and that the Pascal is a very small pressure.

Robert Boyle (1627–91) discovered that the volume V of any fixed mass of gas varies inversely as its pressure p provided the temperature remains constant. This is expressed as:

$$V \propto \frac{1}{p} \qquad [2.12]$$

or

$$pV = \text{Constant} \qquad [2.13]$$

Jacques Charles (1746–1823) discovered the effect of heat on a gas when pressure is kept constant. He found that if a fixed mass of dry gas is kept at constant pressure, the volume is directly proportional to the absolute temperature during its expansion or compression. This can be expressed as

$$V \propto T \qquad [2.14]$$

or

$$\frac{V}{T} = \text{Constant} \qquad [2.15]$$

It is also found that for a fixed mass of gas at constant volume, pressure p is directly proportional to the absolute temperature T. This is called the **Pressure law**, i.e.

$$p \propto T \qquad [2.16]$$

thus

$$p = \text{Constant} \times T \qquad [2.17]$$

and

$$\frac{p}{T} = \text{Constant} \qquad [2.18]$$

These three laws can be combined into one equation called the **ideal gas equation**,

$$\frac{pV}{T} = \text{Constant} \qquad [2.19]$$

Generally speaking, if a fixed mass of gas has values p_1, V_1, and T_1 and some time later the values become p_2, V_2, and T_2, then

$$\frac{p_1 V_1}{T_1} = \frac{p_2 V_2}{T_2} \qquad [2.20]$$

It must be emphasised that in this expression, **absolute temperature** must be used. Example 2.5 illustrates the use of this formula.

Q

Example 2.5

An air compressor contains 100 cm^3 of air at a pressure of 1.0 atmospheres and a temperature of 15 °C. What is the pressure within the pump when the air is compressed to 20 cm^3 at a temperature of 36 °C?

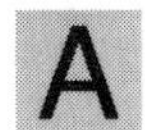

Solution

It should be noted that T is thermodynamic temperature in kelvin (K) and the degrees Celsius scale has to be converted to the kelvin scale.

Thus $T_1 = 273 + 15 = 288$ K

and $T_2 = 273 + 36 = 309$ K

From [2.20]

$$p_1 V_1 / T_1 = p_2 V_2 / T_2$$

$$\frac{1 \times 100}{288} = \frac{p_2 \times 20}{309}$$

therefore

$$p_2 = \frac{100 \times 309}{20 \times 288}$$

$$= 5.36 \text{ atmospheres}$$

Work, energy and power

Work is done when **energy** is changed from one form to another. When an object acted upon by a force of one newton, moves through a distance of one metre, one joule of work is done. This can be expressed as

$$\text{Work done} = \text{Force} \times \text{Distance}$$

or

$$W = Fd \qquad [2.21]$$

The distance moved must be in the direction of the force. A simple example is how much work needs to be done to lift an object weighing 20 kg from the floor, raising it 0.6 m. As the unit of force is the newton (N) and the unit of length is the metre (m), and 1 Nm = 1 J, then since Force (F) = Mass (m) × Gravity (g),

$$W = Fd$$

$$= mgd$$

hence $$= 20 \times 9.81 \times 0.6$$

$$= 117.72 \text{ J}$$

When one joule of work is done in one second, the rate of working is called the **power** of one watt. Power P, is therefore the rate of converting energy and can be expressed as

$$\text{Power} = \frac{\text{Work done}}{\text{Time taken}}$$

$$= \frac{\text{Force} \times \text{Distance}}{\text{Time}}$$

or $$P = \frac{W}{t} \qquad [2.22]$$

Consider the following two examples.

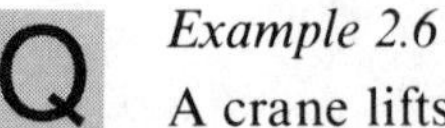

Example 2.6

A crane lifts a load weighing 4000 kg through a height of 12 m in 20 s. What is the power of the crane?

Solution

Here, work done is

$$W = Fd$$

$$= 4000 \times 9.81 \times 12$$

$$= 470\,880 \text{ J}$$

and power is

$$P = \frac{W}{t}$$

$$= \frac{470\,880}{20}$$

$$= 23\,544 \text{ W (or 23.544 kW)}$$

Example 2.7

A motor drives a pump which lifts 100 litres of water at the rate of 4 m/s. What is the power of the motor, ignoring the pump's efficiency?

Solution

In this case the amount of water has to be converted into newtons. Since 1 litre of water has a mass of 1 kg, and 1 kg acting downwards exerts a force of 9.81 N, from [2.22] the power used in 1 s is

$$P = \frac{100 \times 9.81 \times 4}{1}$$

$$= 3924 \text{ W}$$

Note that as
Power = (Force × Distance)/Time
and Distance/Time is velocity, then
Power = Force (N) × Velocity (m/s).

The mechanical power P of a motor is expressed as

$$P = 2\pi nT \qquad [2.23]$$

where n is the motor's **speed** in revolutions per second and T is the motor's **torque** in Nm. Torque is the turning effort and is very similar to the moment of a force as discussed earlier. In fact torque is often called **turning moment**. Figure 2.19 illustrates how torque is developed on a motor's shaft.

Consider a force of F newtons acting tangentially at a point A on a shaft of radius r metres. If the shaft makes one revolution, point A moves through a distance of $2\pi r$ metres and therefore, in one revolution, the force has moved a distance of $2\pi r$ metres.

Hence Work done $W = 2\pi r \times F$ joules

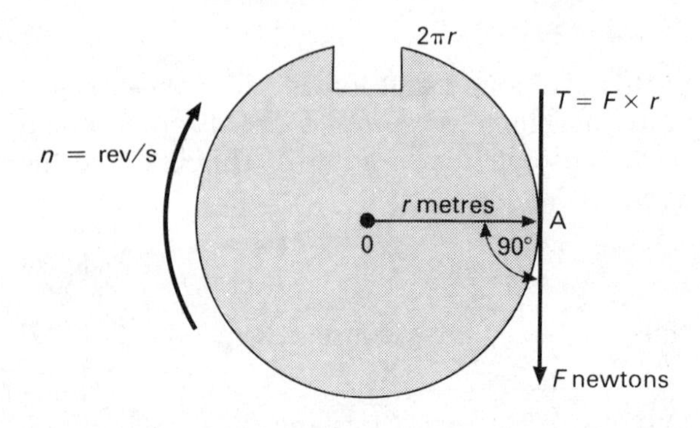

Figure 2.19 Force acting in a circular path

If the shaft rotates at n revolutions per second then

$$\text{Power } P = \text{Work done/Second}$$
$$= 2\pi r \times F \times n$$

and since $\text{Torque } T = Fr$ (Force × Radius)

then $\text{Power } P = 2\pi nT$ watts

Consider the following examples.

Example 2.8

The shaft of a motor is turning at a speed of 1000 rev/min against a torque of 12 Nm. Calculate the motor's shaft power.

Solution

The shaft speed is $1000/60 = 16.667$ rev/s

Since $P = 2\pi nT$

then $P = 2 \times 3.142 \times 16.667 \times 12$

$= 1257$ watts

Example 2.9

A cutting tool in a lathe exerts a tangential force of 80 N on a steel bar of diameter 0.025 m. If the steel bar is turning at 1500 rev/min, what is the power of the lathe?

Solution

From above,

$$T = F \times r = 80 \times 0.0125 = 1 \text{ Nm}$$

Hence

$$\text{Power of the lathe} = 2\pi nT$$
$$= 2 \times 3.142 \times \frac{1500}{60} \times 1$$
$$= 157.1 \text{ W}$$

Simple machines

Levers have already been mentioned, and those such as the crowbar can be described as very simple machines in which a large **load** force can quite easily be moved by a small **effort** force. To see how much the lever is able to magnify the effort force, we use the term **mechanical advantage** (MA). This ratio is expressed as:

$$\text{MA} = \frac{\text{Load}}{\text{Effort}} \qquad [2.24]$$

If a load of 100 N is to be moved and the effort required is only 10 N, then the mechanical advantage is 10. This tells us that, in the case of the crowbar, a person has to push only $\frac{1}{10}$ as hard as the load. Unfortunately, the distance moved by the effort has to be greater than the distance moved by the load. The ratio of these two distances is called the **velocity ratio** (VR). This is expressed as

$$\text{VR} = \frac{\text{Distance moved by the effort}}{\text{Distance moved by the load}} \qquad [2.25]$$

The mechanical advantage depends on how much friction there is in a machine, but friction does not affect the velocity ratio, which refers only to the fact that a smaller force moves a longer distance. In a pulley system, VR equals the number of pulleys used, which equals the number of ropes supporting the load.

Another important term related to machines is **efficiency** η. If you could build a frictionless machine, the work done on the load would equal the work done by the effort. The efficiency of a 'perfect' machine is 100% (a per unit of 1). In general

$$\% \text{ Efficiency} = \frac{\text{Work done on the load}}{\text{Work done by the effort}} \times 100 \qquad [2.26]$$

Since work done is equal to Force × Distance, the output of the machine $(F_1 \times d_1)$ relates to the load to be lifted and the input of the machine $(F_2 \times d_2)$ relates to the effort required. Formulae [2.25] and [2.26] can be expressed as

$$\% \text{ Efficiency} = \frac{\text{MA}}{\text{VR}} \times 100 \qquad [2.27]$$

Example 2.10

A certain machine has an effort force of 20 N and is required to lift a load of 80 N. In order to lift the load by 1 m, the effort needs to be moved 5 m. What is the efficiency of the machine?

Solution

Here,

$$\text{MA} = \frac{\text{Load}}{\text{Effort}}$$
$$= \frac{80}{20} = 4$$

and

$$VR = \frac{\text{Distance moved by effort}}{\text{Distance moved by load}}$$

$$= \frac{5}{1}$$

therefore

$$\% \text{ Efficiency} = \frac{MA \times 100}{VR}$$

$$= \frac{400}{5} = 80\%$$

Example 2.11

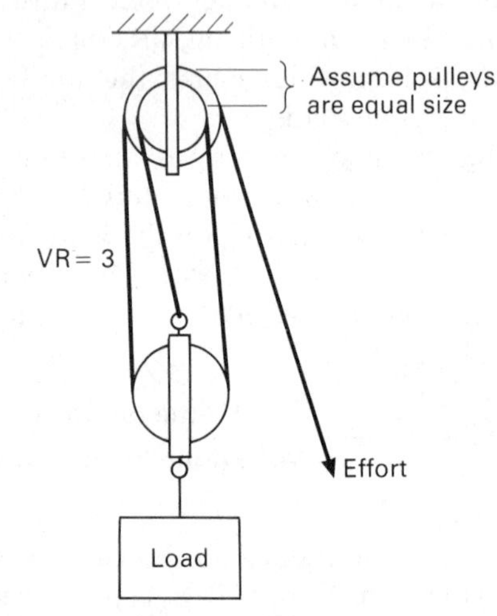

Figure 2.20 Pulley system

Figure 2.20 shows how pulleys are used to lift a load of 600 N with an effort of 250 N. What is the efficiency of the pulley system?

Solution

In this example the VR = 3 since three ropes support the load.

$$MA = \frac{\text{Load}}{\text{Effort}}$$

$$= \frac{600}{250} = 2.4$$

so

$$\% \text{ Efficiency} = \frac{MA \times 100}{VR}$$

$$= \frac{240}{3} = 80\%$$

EXERCISE 2.3

1. A hoist lifts a mass of 15 kg through 10 m at a steady rate in 30 s. Calculate the power output of the hoist.

2. When a force of 12 N is applied to a block of mass 4 kg, it moves along a flat plane at constant velocity (i.e. no acceleration).
   a) What is the force of friction?
   b) What is the resultant force when the force is increased to 30 N?
   c) What is the acceleration in b)?

3. 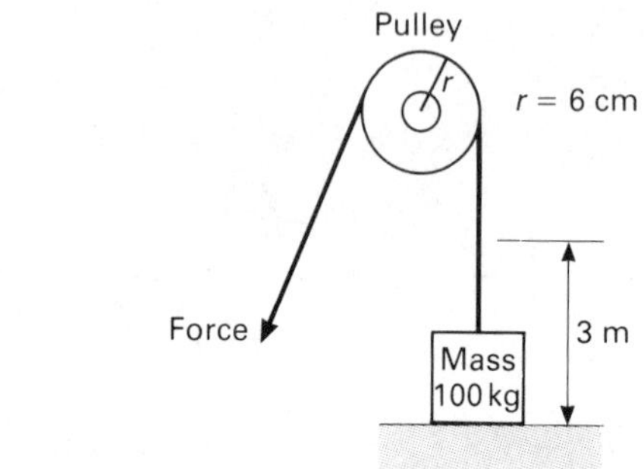

   Figure 2.21 Force in a pulley system

   With reference to Figure 2.21 determine:
   a) the force required to lift the load
   b) the work done in lifting the load 3 m
   c) the torque on the pulley.

4. An electric motor weighing 350 N is slung from a set of blocks which are supported by two wire ropes. The upper ends of the ropes are fixed to a horizontal girder. Find the force in each rope if one is inclined at an angle of 40° and the other at 50°.

5. Determine the force required on the handles of a pair of pliers (see Figure 2.22) if a force of 72 N is required on either side of each cutting edge.

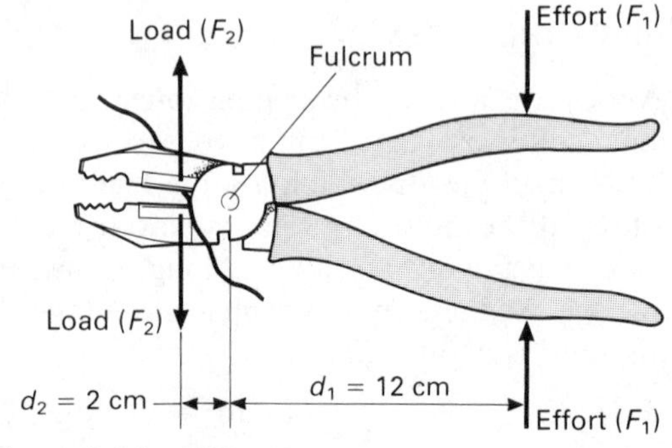

Figure 2.22 Pliers

6. a) A water cylinder has height 0.8 m and area 0.28 m^2. If the density of water is 1000 kg/m^3, what is the mass of water in the cylinder?

b) Given the specific heat capacity of water is 4200 J/kg °C, what heat is required to raise the water temperature from 20 °C to 50 °C?

7. What is the final temperature when 9 m^3 of gas at 60 °C is contracted at constant pressure to 5 m^3?

8. Three rods, aluminium, copper and glass, are all 15 cm long at 20 °C. What will be their lengths at 200 °C if their respective coefficients of linear expansion are: 0.000 026, 0.000 017 and 0 000 009 per °C?

9. a) Determine the torque exerted by a motor-driven machine producing an output of 50 kW at a speed of 1500 rev/min.

b) If the machine has an efficiency of 85%, what does this mean?

10. A pulley system is supported by 4 ropes and is used to lift a load of 300 N. If the efficiency of the system is 75%, what is the effort required to lift the load?

Electrical science

3

Objectives

After reading this chapter you should be able to:

- *describe the structure of the atom in terms of its nucleus and electrons.*
- *describe electric current as a flow of charged particles. State Ohm's Law.*
- *describe resistance, its dependence on dimensions, material and temperature.*
- *describe the connection of ammeters and voltmeters in circuits for the connection of resistors, lamps and other power sources.*
- *describe electricity as a form of energy and also describe power as energy per unit time.*
- *perform simple calculations for finding resistance, current, voltage, power, power factor, energy and efficiency.*
- *describe the heating, magnetic and chemical effects of current.*

Basic circuit theory

Electricity is concerned with the energy present in protons and electrons. It is these and other small particles that constitute the essential ingredients of all atoms, which make up matter in the form of solids, liquids and gases. An atom is not a solid entity, it is often likened to a small solar system having as its centre a nucleus containing positively charged particles called **protons**. Around the nucleus orbit negatively charged particles called **electrons**, which are generally attracted to the nucleus, keeping the atom in a neutral state.

Elements of different substances are given atomic numbers based on how many protons they possess. For example, hydrogen has an atomic number of 1 since it has only 1 proton (and 1 electron); oxygen has an atomic number of 8, copper 29 (see Figure 3.1) and uranium 92.

At times an atom may lose or gain electrons and when this happens it is electrically charged and called an **ion**. If it loses electrons the number of protons is greater than the number of electrons, so the atom is called a **positive ion**. Similarly, if it gains electrons it is called a **negative ion**. The process of

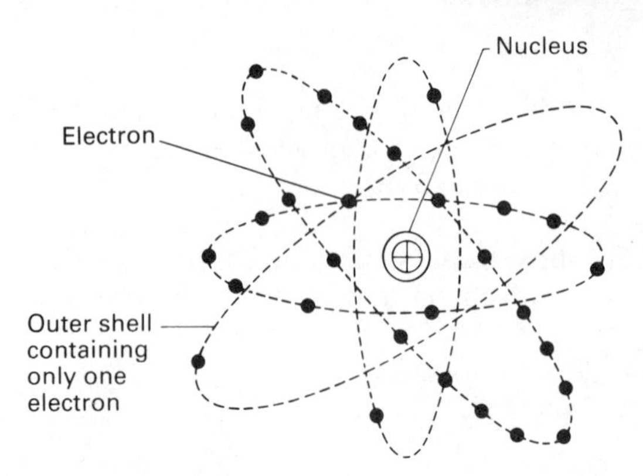

Figure 3.1 Copper atom, showing orbiting negatively charged electrons

forming ions is called **ionisation**, which is a term often used in the operation of gas discharge lamps.

For a material to conduct electricity it must allow charges to flow through it. It is not unusual to find positive and negative charges flowing in opposite directions. In gases and liquid conductors ions can move, whereas in solid conductors only electrons are free to move. Most metals are good conductors of electricity: particularly copper, aluminium and brass. This is because they have ample 'free' electrons available to cause conduction. These electrons flow when they are attracted towards a potential source such as the positive pole of a battery. Figure 3.2 helps to explain this theory.

Materials that cannot conduct electricity within normal temperature limits are those having no free electrons available. Their electrons are completely bonded to the atoms. These materials are called

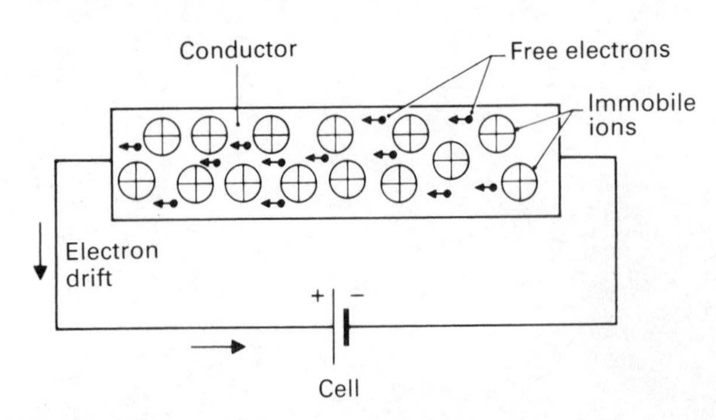

Figure 3.2 Movement of 'free' electrons

insulators. Some examples are rubber, PVC, ebonite, porcelain and glass. Insulators have extremely high resistance values compared with conductors. They act as the medium between 'live' electrical conductors and earth, stopping leakage currents flowing to earth.

Some materials are neither good conductors nor good insulators, as their electrical properties are somewhere between low and high values of resistance. Materials like this, such as germanium and silicon, are called **semiconductors** and are often used as rectifier elements.

The charge of an electron is too small for practical measurement, so a larger unit called a **coulomb** is often used. One coulomb is equivalent to 6.3×10^{18} electrons. This is the **quantity of electricity** (Q) crossing a section of the conductor in a **time** (t) of 1 s and it defines the term **electric current** (I). Expressed as a formula this can be written as

$$I = \frac{Q}{t} \qquad [3.1]$$

The simplest analogy of an electric circuit is to consider a hosepipe connected to a tap. The rate of flow of water from the end of the pipe will depend upon the water pressure at the tap and the flow of water through the pipe, which will be restricted by the inner walls of the pipe, particularly where bends and kinks occur. If there are many restrictions this will be noticeable, as the water will flow out of the pipe at reduced pressure. In the same way, current flows through conductors by means of an electric pressure from a battery or generating source. This source of electric pressure is called **electromotive force** (e.m.f.) and provides the energy to move current through the circuit. Electromotive force is referred to as the supply voltage and for a stable supply the current allowed to flow is determined by resistance in the circuit. There will be a pressure drop across different parts of the circuit and this is called the **potential difference** (p.d.). Unlike the hosepipe analogy, the electric circuit needs a 'go' and 'return' conductor to form a closed loop or circuit, and these must offer a low resistance path to the flow of current. Most solid conductors satisfy this requirement.

Ohm's Law

This law is named after **Georg Ohm** (1787–1854). It states that the ratio of potential difference between the ends of a conductor and the current flowing in the conductor is constant. This ratio is termed the **resistance** of the conductor and is measured in ohms (Ω).

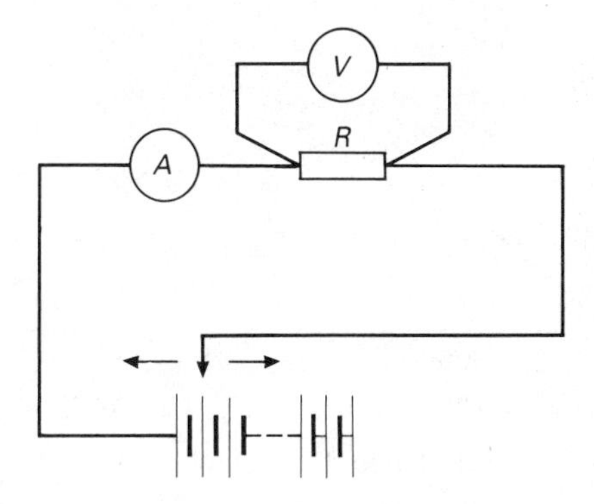

Figure 3.3 Increasing cell voltage increases circuit current

The temperature of the conductor must not change for the law to be obeyed. It can then be stated as:

> The current I flowing in a circuit is directly proportional to the potential difference V and inversely proportional to the circuit resistance R.

Expressed as a formula this statement becomes

$$I = \frac{V}{R} \qquad [3.2]$$

This formula is often used for finding current values such as when selecting suitable fuse sizes and circuit conductors. By rearranging the formula the working resistance of a component, or even the circuit voltage drop, can be found. Figure 3.3 illustrates a method of finding Ohm's Law by varying the supply voltage in a circuit of fixed resistance.

Resistance factors

Experiments have shown that for a particular conductor material at a constant temperature the resistance is:

(i) directly proportional to its length l (see Figure 3.4(a));
(ii) inversely proportional to its cross-sectional area A (see Figure 3.4(b)).

Both length and cross-sectional area are referred to as the conductor's dimensions. Another factor which has to be considered is called **resistivity** (ρ). Some materials conduct electricity more easily than others. Resistivity is a constant for a particular

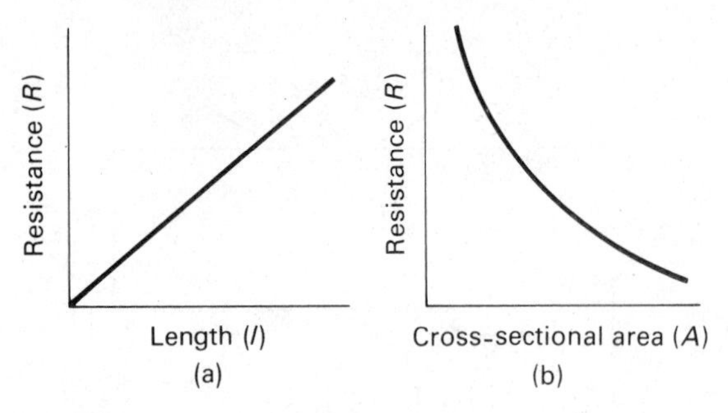

Figure 3.4 Resistance dimensions

material and is defined as the resistance between the opposite faces of a unit cube of the material. It is measured in ohms per metre. Some common values are:

annealed copper	0.0172 μΩ/m
hard-drawn aluminium	0.0285 μΩ/m
rolled brass	0.09 μΩ/m
tungsten	0.056 μΩ/m

These values are measured at a temperature of 20 °C.

The resistance of a conductor is given in terms of its length, cross-sectional area and resistivity as follows

$$R = \frac{\rho l}{A} \qquad [3.3]$$

Note

$$R = \frac{\Omega \cancel{m} \times \cancel{m}}{\cancel{m}^2}$$

Q *Example 3.1*

Determine the resistance of a copper conductor of length 40 m, resistivity 17.2 μΩ/mm and c.s.a. 2.5 mm².

A *Solution*

Since

$$R = \frac{\rho \times l}{A}$$

then

$$R = \frac{0.0172 \times 10^{-6} \times 40}{2.5 \times 10^{-6}}$$

$$= 0.275\ \Omega$$

The resistance of a conductor varies with temperature. Every conductor has a **temperature coefficient of resistance** (α), which relates to the fractional increase per degree Celsius of resistance at 0 °C.

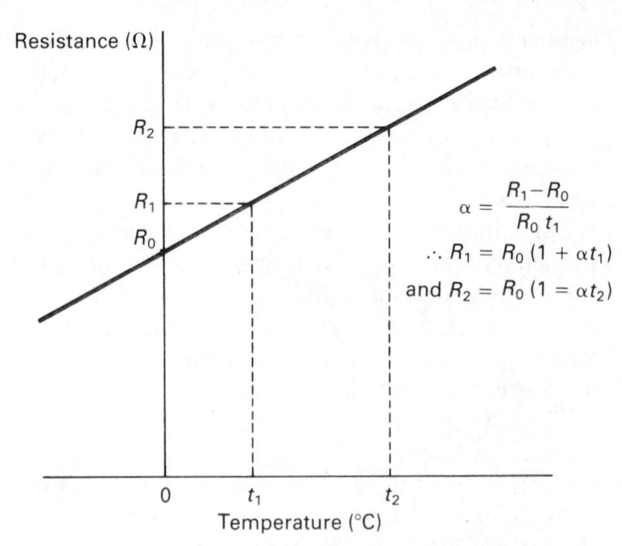

Figure 3.5 Variation of resistance and temperature

This temperature coefficient can be expressed as:

$$\alpha = \frac{R_1 - R_0}{R_0 t_1} \qquad [3.4]$$

Note

$$\frac{\cancel{\Omega}}{\cancel{\Omega} \times {}^\circ\mathrm{C}}$$

where α is the temperature coefficient of resistance (°C^{-1})
R_1 is the resistance at temperature t_1
R_0 is the resistance at 0 °C
t_1 is the temperature (°C).

The resistance of most metal conductors increases as their temperature increases and they are said to have **positive** temperature coefficients of resistance. Most insulators have **negative** temperature coefficients of resistance, including electrolytes, liquids and carbon. Figure 3.5 shows the variation in resistance of a good conductor material such as copper. If the graph of this metal is extended backwards it crosses the base line (i.e. resistance is zero) at a temperature of −235 °C. It is worth noting that copper has a melting point of 1084 °C and it is not too difficult to appreciate that its resistance value will increase considerably over this range of temperature.

Some typical values of α per °C at 0 °C are:

copper	0.004 28 °C^{-1}
aluminium	0.0039 °C^{-1}
carbon	−0.0005 °C^{-1}

Resistor connections

One method of finding the ohmic value of a resistor (i.e. its resistance in ohms) is to connect an

ammeter and **voltmeter** in its circuit, as shown in Figure 3.6. The ammeter directly measures current flowing and should always be placed in series with a load component. The voltmeter directly measures potential difference and should always be connected across components or across live terminals. The working value of resistance is found by dividing the voltmeter reading by the ammeter reading (using Ohm's Law).

If a number of lamps are connected end to end in the form of a chain, they are said to be connected **in series**. In this mode of connection, the same current leaving the supply source passes through each lamp and the sum of all the p.d.s across the lamps will equal the supply voltage. Christmas tree fairy lights are connected like this and although the supply is at 240 V a.c., each lamp requires a voltage of only 12 V. It is very important to realise that if an open circuit occurred in the chain (see Figure 3.7) then the voltage would be 240 V at that point.

If, however, the fairy lights are connected such that each side of a particular lamp shares a common connection with one side of all the other lamps, then they are said to be connected **in parallel**. In this mode of connection each lamp

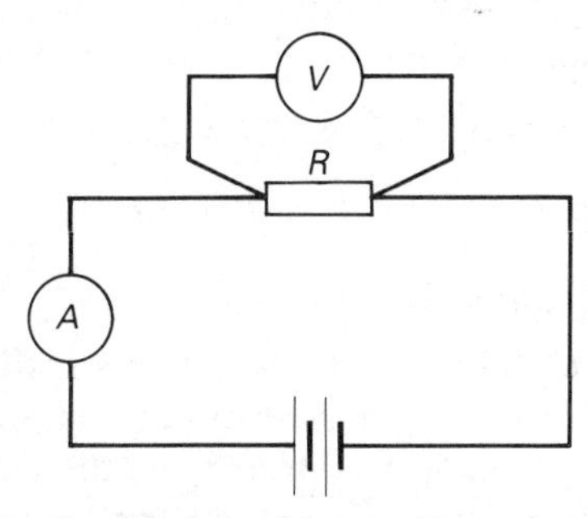

Figure 3.6 Connection of ammeter and voltmeter to determine resistance

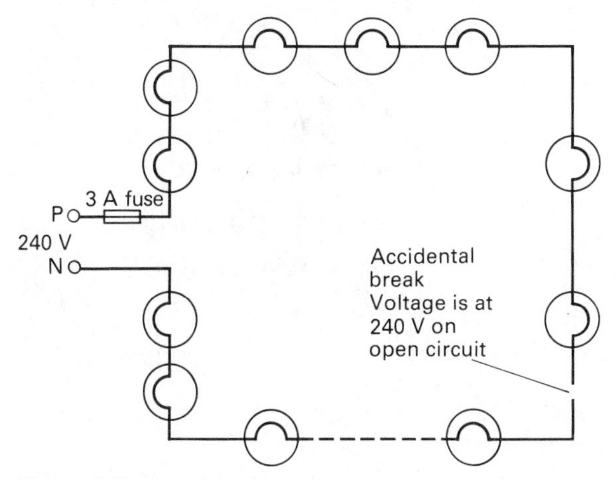

Figure 3.7 Fairy light connections

receives the full supply voltage across it and it therefore needs to be rated the same as the supply voltage, i.e. 240 V.

Another method of connection is called **series-parallel**, which is basically a combination arrangement of the two just described.

Before discussing these methods in detail, it should be pointed out that Ohm's Law applies reasonably well to most parts of the electric circuit. There are also two further laws associated with current distribution and voltage distribution. These are called Kirchhoff's Laws after **Gustav Kirchhoff** (1824–87). The first law states that the total current flowing towards a junction is equal to the current flowing away from that junction; and the second law states that in a closed circuit the p.d.s of each part of the circuit are equal to the resultant e.m.f. in the circuit. These laws will be mentioned in more detail in Part II studies. Parallel resistor circuits are a good example of the first law, while series resistor circuits are a good example of the second law.

RESISTORS IN SERIES

Figure 3.8 shows a typical series circuit where R_1, R_2 and R_3 are three resistors and V_1, V_2 and V_3 are three voltmeters connected across each component measuring the respective p.d.s. The ammeter indicates the current taken from the supply by the three resistors. Applying Ohm's Law to all parts of the circuit:

$$V_1 = IR_1,\ V_2 = IR_2,\ V_3 = IR_3,\ \text{and}\ E = IR_e$$

where E is the e.m.f. or supply voltage and R_e is the equivalent circuit resistance.

Applying Kirchhoff's Second Law

$$E = V_1 + V_2 + V_3$$

$$IR_e = IR_1 + IR_2 + IR_3$$

$$IR_e = I(R_1 + R_2 + R_3)$$

Dividing by I on both sides

$$R_e = R_1 + R_2 + R_3 \qquad [3.5]$$

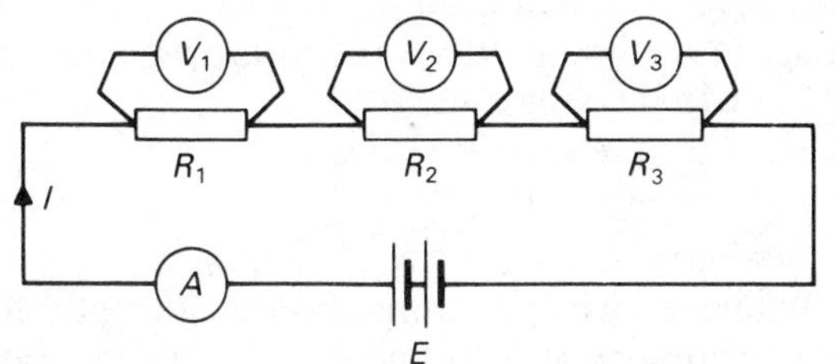

Figure 3.8 Series circuit (common current)

RESISTORS IN PARALLEL

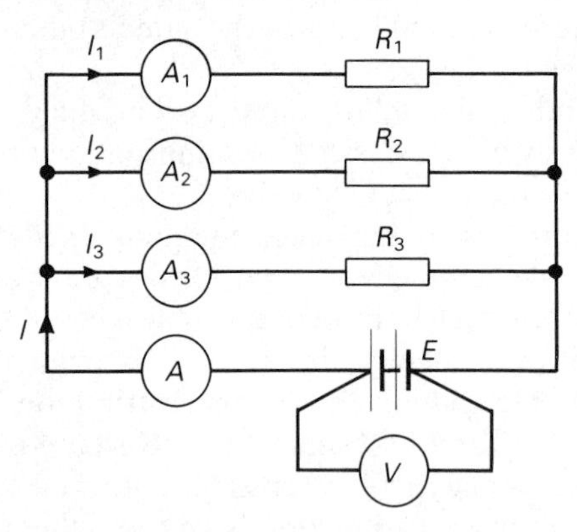

Figure 3.9 Parallel circuit (common voltage)

Figure 3.9 shows a typical parallel circuit using three resistors. There is an ammeter connected in each branch circuit, A_1, A_2 and A_3 and also an ammeter A connected to measure the supply current. Applying Kirchhoff's First Law to the supply current,

$$I = I_1 + I_2 + I_3$$

and according to Ohm's Law

$$I = \frac{E}{R_e}$$

$$\frac{E}{R_e} = \frac{E}{R_1} + \frac{E}{R_2} + \frac{E}{R_3}$$

$$= E\left(\frac{1}{R_1} + \frac{1}{R_2} + \frac{1}{R_3}\right)$$

Dividing by E on both sides

$$\frac{1}{R_e} = \frac{1}{R_1} + \frac{1}{R_2} + \frac{1}{R_3} \qquad [3.6]$$

The point to remember is that $1/R_e$ is the reciprocal of resistance and once it is found it needs to be inverted to $R_e/1$, the equivalent circuit resistance. In a parallel circuit, the value of R_e will always be less than the smallest resistive component in the circuit.

If a parallel circuit comprises only two resistors, then the method to use is called the *product/sum*, which can be written as

$$R_e = \frac{(R_1 \times R_2)}{(R_1 + R_2)}$$

Where a circuit comprises series and parallel connections, treat each parallel circuit separately and find its equivalent resistance. The combination can then be treated as a series circuit, adding the groups together. It is important not to confuse E and V. Remember that V is a p.d. and will be less than E, the e.m.f.

Q

Example 3.2

Three resistors, 1 Ω, 2 Ω and 3 Ω are connected in series across a 12 V battery. Find the equivalent circuit resistance, the current taken from the supply and the p.d. across each resistor.

Solution

$$R_e = R_1 + R_2 + R_3$$

$$= 1 + 2 + 3$$

$$= 6\ \Omega$$

$$I = \frac{E}{R_e}$$

$$= \frac{12}{6}$$

$$= 2\ \text{A}$$

Across R_1 p.d. is $V_1 = IR_1 = 2 \times 1 = 2$ V

Across R_2 p.d. is $V_2 = IR_2 = 2 \times 2 = 4$ V

Across R_3 p.d. is $V_3 = IR_3 = 2 \times 3 = 6$ V

Thus $E = IR_1 + IR_2 + IR_3$

Example 3.3

Consider the three resistors in Example 3.2 now connected in parallel across the same supply source. Find the equivalent circuit resistance, the supply current and the branch currents in each resistor.

Solution

$$\frac{1}{R_e} = \frac{1}{R_1} + \frac{1}{R_2} + \frac{1}{R_3}$$

$$= 1 + 0.5 + 0.333$$

$$= 1.833\ \Omega$$

thus $\quad R_e = 0.545\ \Omega$

$$I = \frac{E}{R_e}$$

$$= \frac{12}{0.545}$$

$$= 22\ \text{A}$$

Current in R_1 is $I_1 = E/R_1 = 12/1 = 12$ A
Current in R_2 is $I_2 = E/R_2 = 12/2 = 6$ A
Current in R_3 is $I_3 = E/R_3 = 12/3 = 4$ A
Thus $I = I_1 + I_2 + I_3$

It should be noted that both these circuits have identical components, yet the parallel circuit equivalent resistance is eleven times smaller than the series circuit. In real terms the parallel circuit will dissipate eleven times more heat. Resistors in series create voltage drops and their main purpose is to block current flow.

Example 3.4

Two 12 Ω resistors are connected in parallel across a 12 V battery. What is their equivalent resistance and the current taken from the battery?

Solution

$$R_e = \frac{(R_1 \times R_2)}{(R_1 + R_2)} = \frac{(12 \times 12)}{(12 + 12)} = 6\ \Omega$$

$$I = \frac{E}{R_e} = \frac{12}{6} = 2\text{ A}$$

You should note that where two or more **identical** resistors are connected in parallel, the equivalent resistance is divided by the number of resistors. For example, five 20 Ω resistors would be equivalent to a resistance of 4 Ω.

Example 3.5

Figure 3.10 shows a series-parallel combination of resistors. Determine the unknown resistance R_2 if the equivalent circuit resistance is 5 Ω.

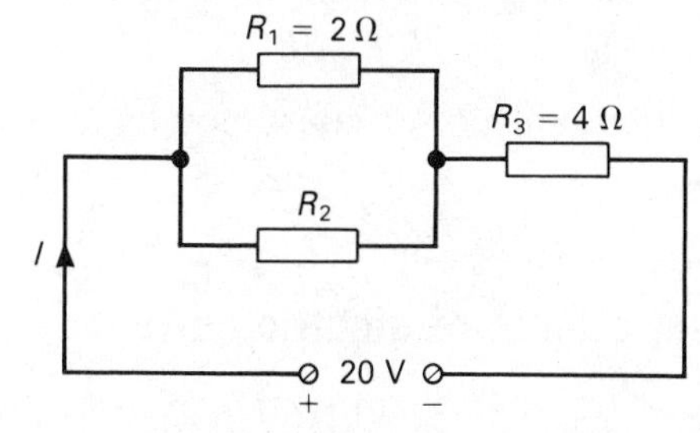

Figure 3.10 Series-parallel resistors

Solution

Supply current is

$$I = \frac{E}{R_e} = \frac{20}{5} = 4\text{ A}$$

P.d. across 4 Ω resistor

$$V_3 = I \times R_3 = 4 \times 4 = 16\text{ V}$$

Therefore p.d. across parallel resistors is 4 V.

Current in 2 Ω resistor is

$$I = \frac{V}{R_1} = \frac{4}{2} = 2\text{ A}$$

Thus 2 A must flow in resistor R_2 which must therefore also have resistance 2 Ω.

Example 3.6

A series-parallel combination of lamps is shown in Figure 3.11. Determine the equivalent circuit resistance, supply current and p.d. across each part of the circuit.

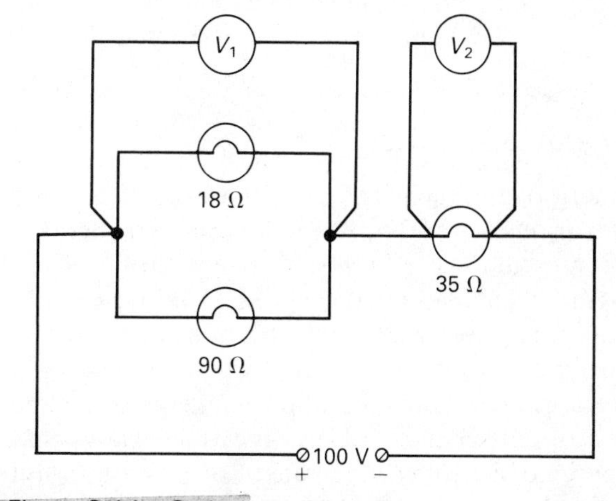

Figure 3.11 Series-parallel lamps

Solution

For the parallel group

$$R = \frac{(R_1 \times R_2)}{(R_1 + R_2)}$$

$$= \frac{(90 \times 18)}{(90 + 18)}$$

$$= 15\ \Omega$$

Equivalent resistance

$$R_e = 15 + 35 = 50\ \Omega$$

Circuit current

$$I = \frac{E}{R_e}$$

$$= \frac{100}{50}$$

$$= 2\ \text{A}$$

P.d. across 15 Ω resistor

$$V = I \times R$$

$$= 2 \times 15$$

$$= 30\ \text{V}$$

P.d. across 35 Ω resistor

$$V = I \times R$$

$$= 2 \times 35$$

$$= 70\ \text{V}$$

Note: The current through each branch circuit of the parallel connected lamps is found by $I = V/R$. For the 18 Ω lamp, $I = 30/18 = 1.67$ A.

The 90 Ω lamp takes a current of $I = V/R = 30/90 = 0.33$ A and the 35 Ω lamp takes $I = 70/35 = 2$ A.

Energy and power

Energy is the capacity for doing work and it exists in various forms. For example, **potential energy** is energy that a body possesses by virtue of its position, such as a coiled spring. Another form of energy is **kinetic energy**, which a body possesses by virtue of its motion, such as when the spring is released. This transformation of energy obeys the law of conservation of energy, which states that energy can neither be created nor destroyed, but only changed from one form to another.

The generation of electricity is a good example of energy conversion. In power stations, fossil fuels such as coal or oil are burned to produce heat. This heat is used to boil water, converting it into high pressure steam, which drives large turbines. These turbines are mechanically coupled to a.c. generators, which produce the electricity required. The conversion cycle actually passes through four main stages, namely chemical energy, heat energy, mechanical energy and electrical energy. All these forms of energy have the same basic unit of measurement, which is called the **joule** (J).

In an electric circuit, the voltage source (known as the **electromotive force** E) is measured in terms of the **energy** (W), or number of joules of work necessary to move a **quantity of electricity** (Q) around the circuit. This can be expressed as

$$E = \frac{W}{Q} \qquad [3.7]$$

By transposition

$$W = EQ$$

Since in [3.1] $\quad Q = It$

we have $\quad W = EIt \qquad [3.8]$

Power (P) is the rate of doing work, i.e. it is **energy per unit time**. Its unit is the watt, named after **James Watt** (1736–1819), and defined as a rate of working of one joule per second. This ratio can be expressed as

$$P = \frac{W}{t} \qquad [3.9]$$

By rearranging the formula

$$W = Pt$$

but in formula [3.8]

$$W = EIt$$

so combining [3.8] and [3.9] we obtain

$$Pt = EIt$$

Thus $\quad P = EI \qquad [3.10]$

This is the total power consumed by an electric circuit.

Heating effects of electric current

When a current flows through a resistor, heat is dissipated between the points of its connection.

The power consumed in this part of the circuit involves both current (I) and potential difference (V) and not the electromotive force (E). In this case, formula [3.10] can be written as

$$P = VI \qquad [3.11]$$

There are several ways of expressing power. For example, in the formula for Ohm's Law [3.2]

$$I = \frac{V}{R}$$

but substituting I in [3.11] gives

$$P = \frac{VV}{R}$$

$$= \frac{V^2}{R} \qquad [3.12]$$

Also, since $V = IR$,

$$P = IRI$$

$$= I^2R \qquad [3.13]$$

The instrument that records energy is called an **integrating meter** or **energy meter** and the instrument that measures power is called a **wattmeter**. They should not be confused, since the former records time. Both types can be used on a.c. and d.c. supplies. Figure 3.12 shows their connection in a single-phase a.c. circuit.

When reading consumers' energy meters, you will come across dial-type meters as well as digital types, which are a later development gaining widespread use, particularly in conjunction with Economy 7 metering arrangements. You should know that 1000 watt hours (1 kWh) is a unit of electricity and Electricity Boards make unit charges to consumers based on their consumption of electricity over a quarterly period. A further standing charge is made to cover meter-reading costs, maintenance costs, equipment costs, etc. For further information on reading electricity meters and an understanding of retail tariffs, see Book 2 in this series. Sometimes the energy meter can be used to find the rating of a piece of equipment by using the meter's disc constant, which is 240 revolutions per kilowatt hour or per unit. From formula [3.9] $P = W/t$. The rating of the equipment is

$$P = \frac{\text{No. of Revs}}{(\text{Time taken} \times 240)}$$

For example, if the equipment is timed for 5 minutes (0.083 hrs) and 40 revolutions are counted as the disc rotates, then

$$P = \frac{40}{(0.083 \times 240)} = 2\ \text{kW}$$

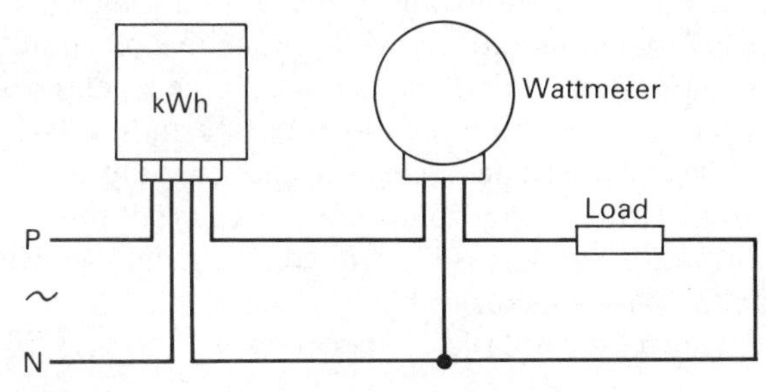

Figure 3.12 Energy meter and wattmeter connections

Q

Example 3.7

Two resistors of 12 Ω and 28 Ω respectively are connected in series and then in parallel across a 200 V supply. Which mode of connection will produce the most heat?

A

Solution

In series $R = 40\ \Omega$

and $I = \frac{V}{R} = \frac{200}{40} = 5\ \text{A}$

therefore $P = V \times I = 200 \times 5 = 1000\ \text{W}$

In parallel $R = 8.4\ \Omega$

and $I = \frac{V}{R} = \frac{200}{8.4} = 23.81\ \text{A}$

therefore $P = V \times I = 200 \times 23.81 = 4762\ \text{W}$

The parallel resistors produce the most heat.

Q

Example 3.8

The estimated daily loading of a consumer's final circuits is:

lighting	0.5 kW for 6 hours
water heating	3.0 kW for 2 hours
ring circuit	3.5 kW for 3 hours
cooking	5.0 kW for 2 hours

Determine the energy used daily, in
a) kilowatt hours b) megajoules.

Note: 1 kWh = 3.6 MJ

A

Solution

a) $W = Pt$

$$= (500 \times 6) + (3000 \times 2) + (3500 \times 3) + (5000 \times 2)$$

$$= 29\,500 \text{ Wh}$$

$$= 29.5 \text{ kWh}$$

b) Since $1 \text{ kWh} = 3.6 \text{ MJ}$

then $W = 29.5 \times 3.6$

$$= 106.2 \text{ MJ}$$

Note: If the cost of energy is 6.12 p/unit (kWh) then the daily charge is $29.5 \times 6.12 = 180.54$p or £1.80.

Q

Example 3.9

An electric kettle is rated at 3 kW/240 V. Find the cost of using the kettle 28 times per week if it takes 2 minutes to boil. Assume one unit of electricity costs 5.8p.

A

Solution

Since $W = Pt$ in kWh,

then $W = \frac{3 \times 2}{60} \times 28$

$$= 2.8 \text{ kWh}$$

therefore

Cost per week $= 2.8 \times 5.8$

$$= 16\text{p}$$

Note: Over one quarter (13 weeks), the cost would be £2.11.

Q

Example 3.10

A socket outlet is fed by PVC cables in steel conduit from a 240 V a.c. supply.

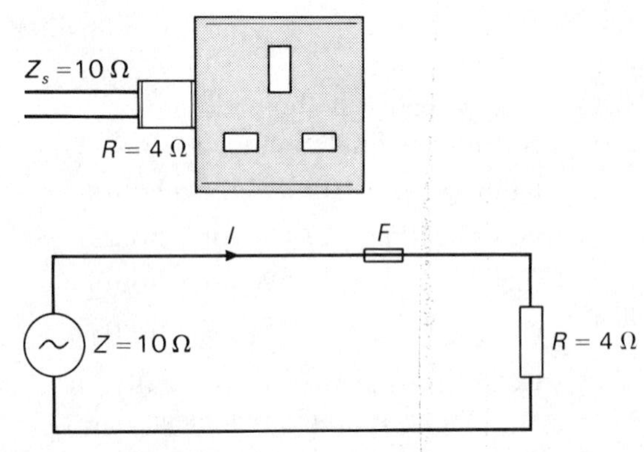

Figure 3.13 High resistance joint

The phase conductor develops a fault of negligible impedance to earth at the socket. If the earth fault loop impedance from the socket is 10 Ω, determine the fault current and power consumed by a joint in the conduit if its resistance is 4 Ω. (See Figure 3.13.)

A

Solution

$$I = \frac{V}{Z} = \frac{240}{10}$$

$$= 24 \text{ A}$$

$$P = I^2R = 24 \times 24 \times 4$$

$$= 2304 \text{ W } (2.304 \text{ kW})$$

Note: In this problem the term impedance (Z) combines the effects of both resistance (R) and reactance (X) encountered by the earth fault current as it travels through the steel conduit and supply transformer, winding back to the point of fault again. See Chapter 1 for a definition of these terms.
If the socket outlet were wired as part of a ring final circuit, it would be protected by a 30 A overcurrent device. Under these circumstances a fault current of 24 A would not be sufficient to cause the device to disconnect the circuit and considerable heat would be generated at the joint.

Power factor

It was shown in formula [3.11] that $P = VI$, but in a.c. circuits loads may not always be resistive: they may possess **inductance** (L) and have **inductive reactance** (X_L) or they may possess **capacitance** (C) and have **capacitive reactance** (X_c). Inductive reactance is found in a motor winding or a discharge lamp ballast, while capacitive reactance is found in a bank of capacitors. These reactances will be dealt with more fully in Part II studies. They cause a **phase angle** (ϕ) to occur between the circuit current and supply voltage and because of this the formula for power is multiplied by a factor known as the **power factor** (p.f.).

Hence, formula [3.11] becomes

$$P = V \times I \times \text{p.f.} \qquad [3.14]$$

By transposition

$$\text{p.f.} = \frac{P}{(V \times I)}$$

The power factor is therefore the ratio between active power (P) and apparent power (VI). The active power is measured by a wattmeter whereas the apparent power is measured by a voltmeter and ammeter. Figure 3.14 shows how these three instruments are connected to determine the power factor of a circuit. Single instruments known as power factor indicators are available and suitable for connection on either single phase a.c. or three phase a.c. supplies.

Figure 3.15 shows the typical circuit connections of a **dynamometer type** p.f. indicator, which has a 90° movement. Moving iron types are also made, which have a 360° movement. In terms of operation, if a 'purely' resistive load were connected in a circuit the instrument's pointer would indicate **unity power factor** because there is no phase angle between the supply current and supply voltage. Unity power factor is 1 and whenever a phase angle occurs (through the connection of an inductor or capacitor) the power factor worsens and becomes less than unity (a decimal). If an inductive load (e.g. choke or winding) were connected then the instrument would read a lagging power factor. If the load were capacitive (e.g. a capacitor) the instrument would read a leading power factor. Power factor may therefore be considered to be in any one of three states: either unity, lagging or leading. It is the lagging state that affects us most, as it causes the current quantity to lag behind the voltage quantity by an undesirable phase angle. This in turn causes more current than necessary to be taken into the circuit, leading to more expensive cabling and equipment costs.

Example 3.11 illustrates different power factor conditions. Note which one creates the minimum current flow in the circuit.

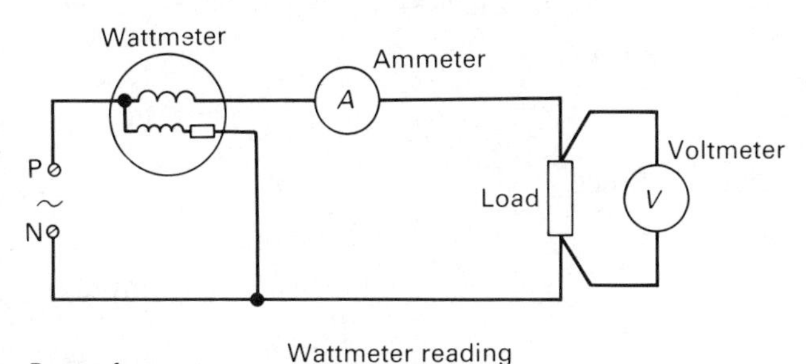

Figure 3.14 Instrument connections

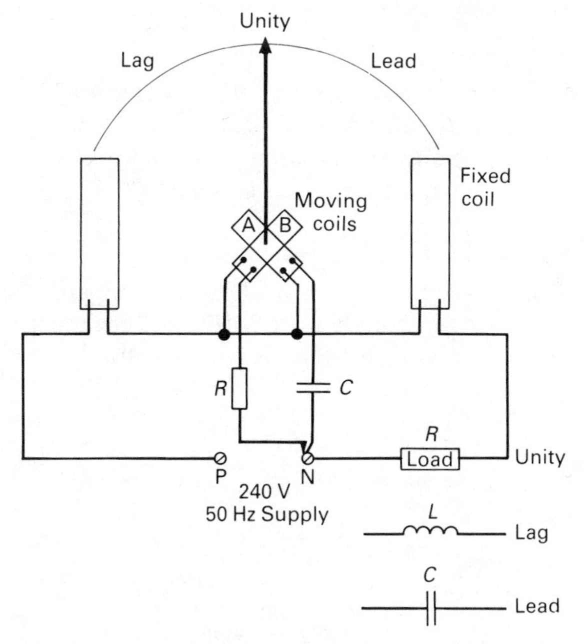

Figure 3.15 Dynamometer type p.f. indicator

Example 3.11

What current is taken by a 3 kW/240 V single phase a.c. load if its power factor changes between

a) 0.2 leading;
b) 0.8 leading;
c) unity;
d) 0.7 lagging;
e) 0.2 lagging?

Solution

From formula [3.14]

$$P = V \times I \times \text{p.f.}$$

a) $$I = \frac{P}{(V \times \text{p.f.})}$$

$$= \frac{3000}{(240 \times 0.2)} = 62.5\ \text{A}$$

b) $$I = \frac{3000}{(240 \times 0.8)} = 15.62\ \text{A}$$

c) $$I = \frac{3000}{(240 \times 1)} = 12.5\ \text{A}$$

d) $$I = \frac{3000}{(240 \times 0.7)} = 17.85\ \text{A}$$

e) $$I = \frac{3000}{(240 \times 0.2)} = 62.5\ \text{A}$$

Efficiency

The efficiency (η) of a system is the ratio of work output to work input. If a machine were capable of being 100 % efficient it would have no losses and its efficiency would be 1. In general, useful work put into a system will always be greater than the work output. Efficiency is often expressed as a percentage or in per unit form and the ratio must always be in the same units. Thus:

$$\%\ \text{Efficiency} = \frac{\text{Work output} \times 100}{\text{Work input}}$$

Expressed as a formula

$$\eta = \frac{(P_O \times 100)}{P_I} \qquad [3.15]$$

Q

Example 3.12

Determine the efficiency of a water heating system if the heat energy input is 800 MJ and the heat output is 500 MJ.

A

Solution

$$\%\ \text{Efficiency}\ (\eta) = \frac{\text{Work output} \times 100}{\text{Work input}}$$

$$= \frac{(500 \times 100)}{800}$$

$$= 62.5\%$$

Note: In this example, 37.5% of heat is being lost and ways of making the heating system more efficient should be considered. What would you suggest?

Taking the example above a little further, the amount of heat energy required to raise the temperature of a mass of 1 kg of water by 1 °C (1 K) is approximately 4200 J. Heat energy output (Q) is directly proportional to the mass (m) of the substance and the temperature rise (or difference in initial temperature (t_1) and final temperature (t_2)). This can be expressed as

$$Q = mc(t_2 - t_1)\ \text{J} \qquad [3.16]$$

(see page 24). Consider Example 3.13.

Q

Example 3.13

A tank containing 1300 litres of water is heated from an initial temperature of 15 °C to 55 °C. If the electrical supply is a 15 kW heater operated for a period of 10 hours, what is the efficiency of the system?

A

Solution

Here

$$Q = 1300 \times 4200 \times 40 = 218.4\ \text{MJ}$$

Since

$$3.6\ \text{MJ} = 1\ \text{kWh},$$

$$Q = 60.67\ \text{kWh}$$

The electrical energy input required to heat the water for 10 hours is given by the formula $W = Pt$

Hence

$$W = 15 \times 10 = 150\ \text{kWh}$$

Since

$$\%\ \text{Efficiency} = \frac{\text{Heat output} \times 100}{\text{Electrical energy input}}$$

then

$$\eta = \frac{Q}{W} = \frac{60.67 \times 100}{150}$$

$$= 40.45\%$$

Q

Example 3.14

A 240 V a.c. motor takes a current of 47.62 A and has a power factor of 0.7 lagging. If its output rating is 6.5 kW, what is the motor's efficiency?

Solution

Here the mechanical output rating of the motor (P_O) is 6.5 kW. The electrical input is given by formula [3.14]
$P_I = V \times I \times \text{p.f.}$
Thus

$$P_I = 240 \times 47.62 \times 0.7 = 8\ \text{kW}$$

$$\eta = \frac{P_0 \times 100}{P_I}$$

$$= \frac{6.5 \times 100}{8} = 81.25\%$$

Example 3.15

Figure 3.16 shows a motor driving a machine through its shaft. If the motor has an efficiency of 85% and the machine has an efficiency of 75%, what is the rating of the motor if the power required by the machine is 4 kW?

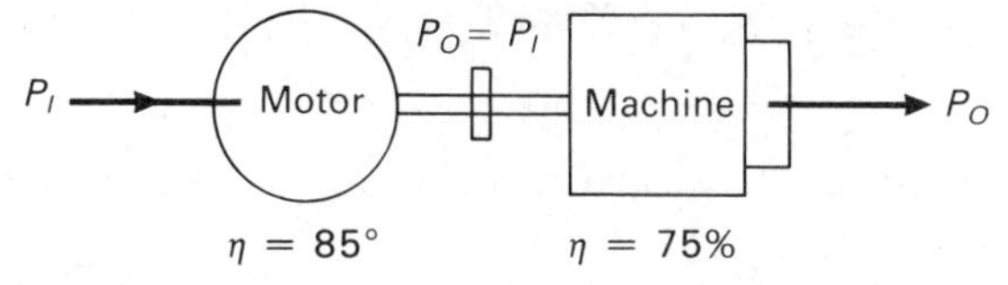

Figure 3.16 Motor and machine efficiencies

Solution

In this question the shaft of the motor is its output, which is common to the machine's input.

Transposing formula [3.15] the machine's input is

$$P_I = \frac{P_O \times 100}{\eta} = \frac{400}{75} = 5.33 \text{ kW}$$

Since the motor is 85% efficient and its P_O is 5.33 kW then

$$P_I = \frac{P_O \times 100}{\eta} = \frac{533}{85} = 6.27 \text{ kW}$$

Magnetic effects of electric current

Hans Christian Oersted (1777–1854) discovered that passing a direct current through a copper wire made a compass needle held at the side of the wire swing to one side. When he reversed the current direction the compass needle swung to the opposite side. Oersted had demonstrated that an electric current has a magnetic effect.

The current-carrying conductor produces a magnetic field, which can be made visible by using iron filings sprinkled on a flat card with the conductor passing through the centre of it. By placing the compass needle on the card the direction of the field can be traced (see Figure 3.17). It will be seen that these lines of force are concentric circles. If current is flowing inwards or down through the card, the lines of magnetic force follow a clockwise direction; if current is flowing outwards or up through the card, the magnetic lines follow an anticlockwise direction. This is illustrated using the dot and cross notation shown

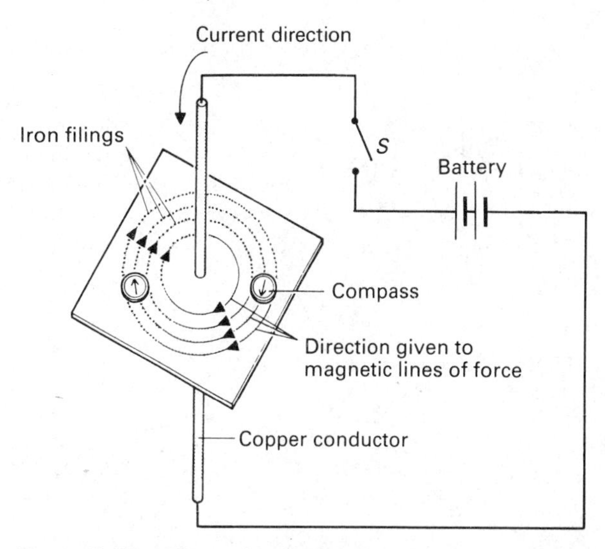

Figure 3.17 Magnetic effects of current

in Figure 3.18(a). It gives both current direction and magnetic field direction. A more practical way to remember this is by using the Corkscrew Rule, which says that if you screw a corkscrew (or ordinary screw) in the direction of current, its rotation indicates the magnetic flux direction of a free north pole (see Figure 3.18(b)).

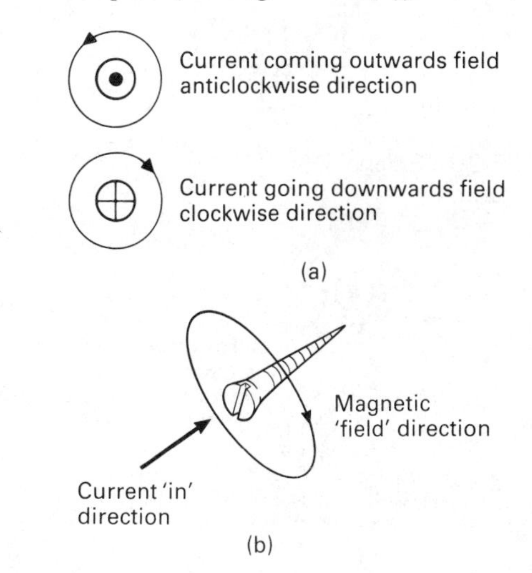

Figure 3.18 (a) Dot and cross rule (b) Corkscrew Rule

Another method of finding the direction of magnetic flux is to use the right hand grip rule (see Figure 3.19), where the curl of your fingers represents the direction of current and your thumb points in the direction of magnetic field. This rule is particularly useful in determining the field pattern set up around a **solenoid**, see Figure 3.20. Further examples using the dot and cross notation are given in Figure 3.21. It is worth mentioning that

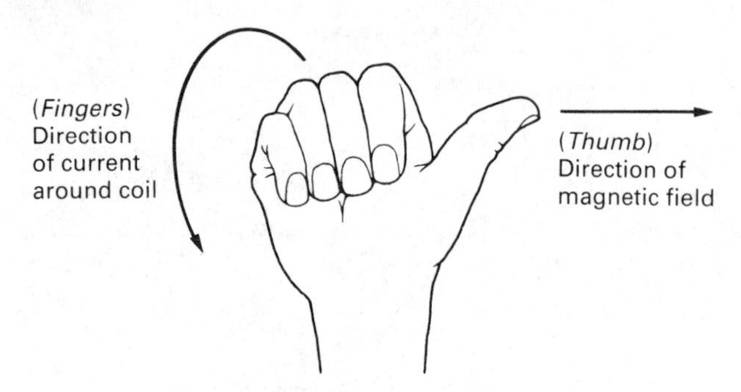

Figure 3.19 Grip rule

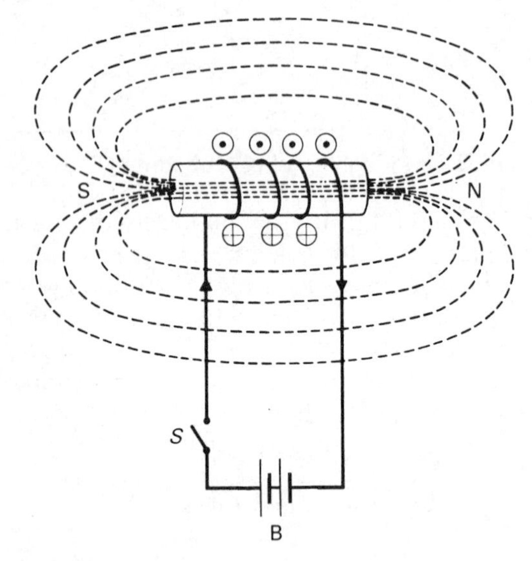

Figure 3.20 Solenoid

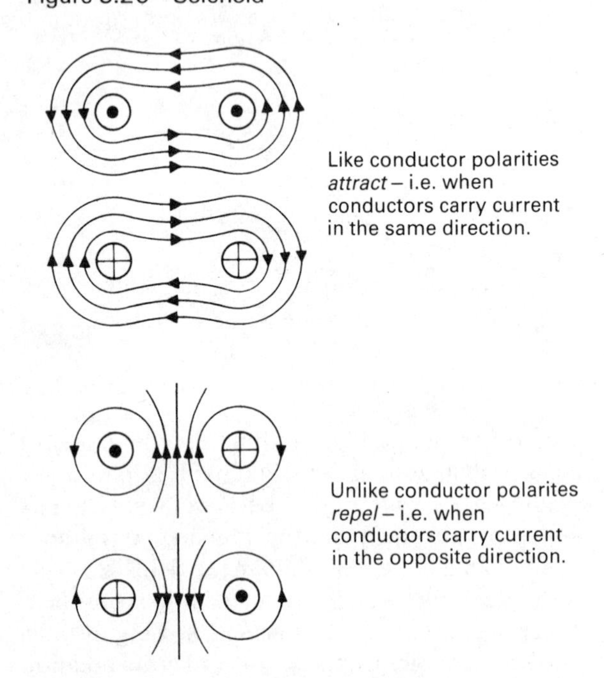

Figure 3.21 Magnetic field behaviour

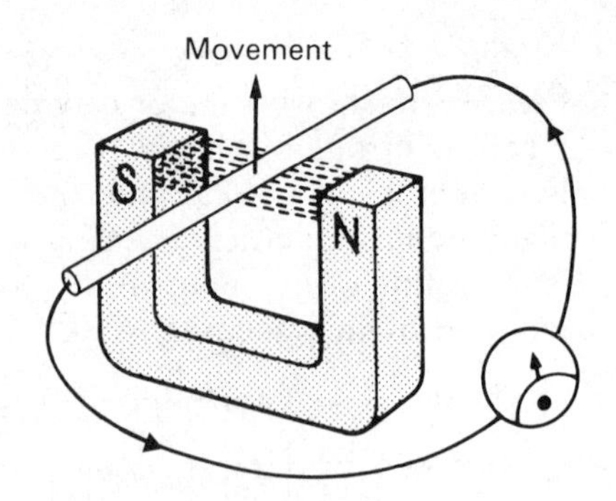

Figure 3.22 Copper conductor and stationary magnet

these magnetic flux lines obey certain rules. For example, they each form a closed loop, always try to contract, especially when stretched, and never intersect each other. Moreover, when they are parallel and in the same direction they attract one another.

It was **Michael Faraday** (1791–1867) who discovered the principles of **electromagnetic induction**. He found that when a bar magnet was moved towards a stationary coil, the magnet's field created electricity in the coil. By moving the bar magnet away from the coil, he could make the electricity flow in the opposite direction. This principle is the same if the coil is moved instead of the magnet. Figure 3.22 shows a short piece of copper conductor being moved inside a stationary permanent magnet. The copper wire is connected to a sensitive galvanometer. The galvanometer needle will deflect in one direction and then in the opposite direction when the wire is passed in and out of the field. The faster this is done, the greater will be the deflection of the needle. The induced e.m.f. is equal to the average rate of cutting magnetic flux. Thus

$$E = \frac{\Phi}{t} \qquad [3.17]$$

where E is the e.m.f. induced in volts (V)
Φ is the magnetic flux in webers (Wb)
t is the time in seconds (s)

There are three factors which alter the strength of an induced e.m.f.

1. The strength of the magnetic flux density (B) between the magnetic poles.
2. The length (l) of conductor in the magnetic field.
3. The velocity (v) or speed of the conductor passing through the magnetic field.

Hence $$E = Blv \qquad [3.18]$$

John Fleming (1849–1945) devised rules for finding the direction of induced e.m.f., magnetic field and motion – a right hand rule for generators and a left hand rule for motors. Figure 3.23 shows how the fingers and thumb of the right hand are kept at right angles to each other to indicate these quantities. You can apply this rule to Figure 3.25, which shows how a voltage is created in a rotating loop as it passes between the poles of a permanent magnet. You must accept the convention that magnetic lines of force emanate from a north-seeking pole and enter a south-seeking pole.

Heinrich Lenz (1804–65) discovered that when a circuit and a magnetic field move relative to each other the direction of induced e.m.f. is always such that it sets up a current opposing the motion or the change in flux responsible for inducing that e.m.f. This is illustrated in Figure 3.24 where it will be seen that by moving the permanent magnet closer to the coil a similar pole is created at the end of the coil. This tells us that the current is opposing the movement of the magnet by trying to repel it. Moving the magnet away makes the current in the coil move the other way to produce an opposite pole, which again opposes the movement of the magnet by trying to attract it.

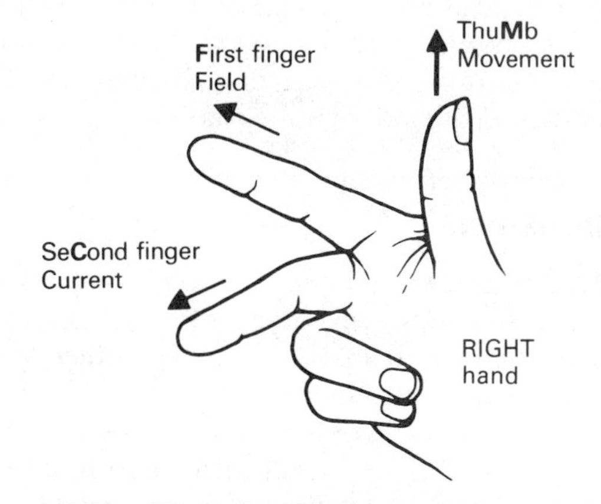

Figure 3.23 Fleming's right hand rule

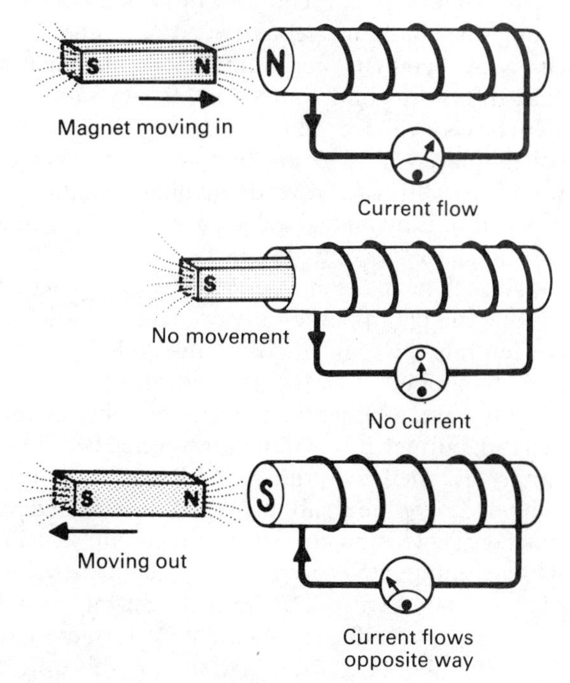

Figure 3.24 Electromagnetic induction

At this stage, you have seen that current flow is a result of some form of electromotive force. In practice, a distinction is made between two main sources of supply, namely **direct current** (d.c.) and **alternating current** (a.c.). The former can be created chemically by cells or batteries or even from d.c. generators such as a dynamo, whereas a.c. is created by a.c. generators known as alternators. Both a.c. and d.c. generators operate on the principle of electromagnetic induction, but it is the a.c. type which is used for public supplies into consumers' premises. It can quite easily be rectified to d.c., as well as being transformed to a higher or lower voltage, using **transformers**.

Simple loop generators

Figure 3.25 shows a diagram of a simple single loop generator. The d.c. machine has a part on it called a **commutator** so that it can draw a unidirectional e.m.f. as its loop rotates. The current is drawn through the brushes resting on the commutator. When the loop is externally rotated there will be maximum e.m.f. induced in the position shown, i.e. under the main poles. As the loop moves further around, less e.m.f. will be induced, and no e.m.f. at all when the loop is travelling perpendicular to the main field path. It is at this point in the rotation that the brushes come into contact with the opposite commutator segments. Thus, although the direction of the current induced in the coil is reversed, the external current will be in the same direction as previously. The induced e.m.f. begins to rise again as the loop moves to a horizontal position between the main poles again. The external current will rise to a maximum and then fall to zero on each half cycle of 180°. This will be repeated as shown on the graph. The more commutator loops there are, the smoother will be the d.c. output.

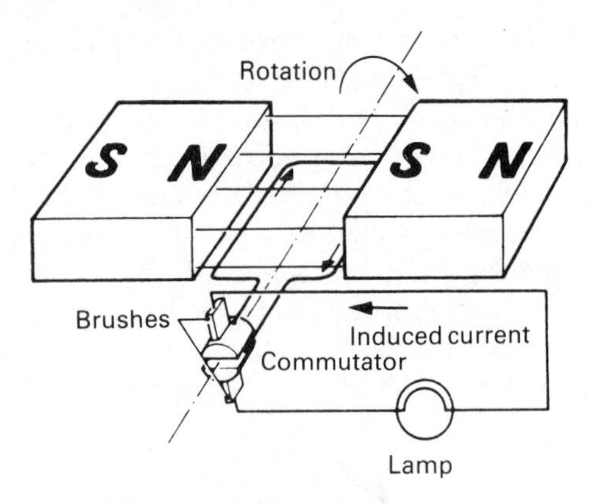

(a) Single loop generator showing commutator

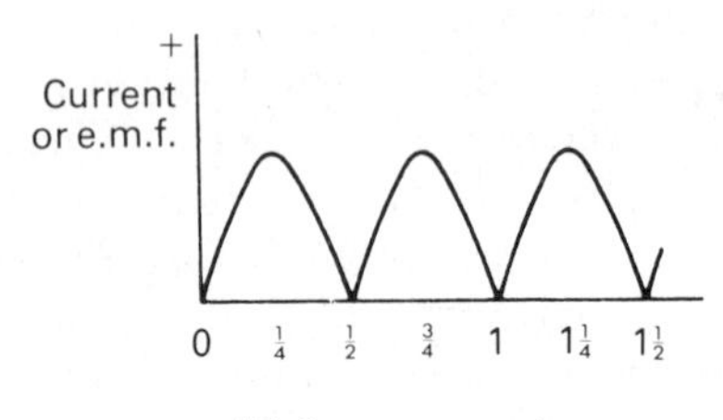

(b) Output waveshape

Figure 3.25 Generation of direct current

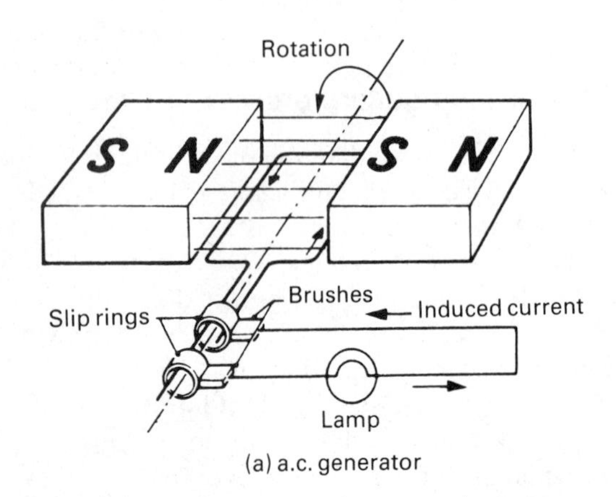

(a) a.c. generator

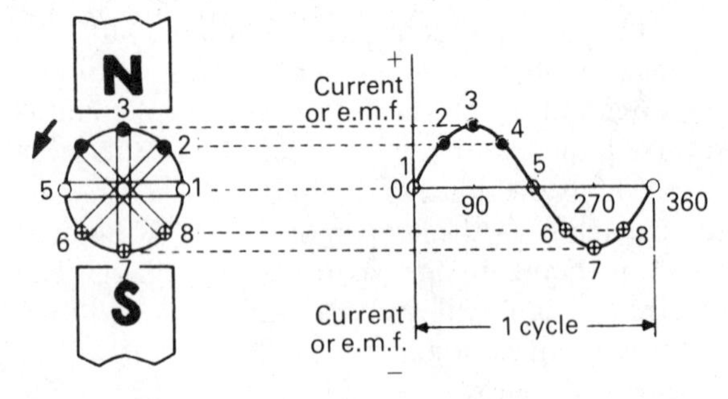

(b) Output waveshape

Figure 3.26 Generation of alternating current

Figure 3.26 shows how a single loop a.c. generator operates. Unlike the d.c. generator the loop is attached to **slip rings** but again brushes are used to supply an external circuit. When the loop is externally rotated, maximum flux cutting occurs underneath the main poles (positions 3 and 7 on the graph). In positions 1 and 5 the loop is travelling perpendicular to the field and no induced e.m.f. occurs. Since the slip rings are not segmented the output supply will change direction as the e.m.f. induced in the loop changes direction. It will be cyclic over one complete revolution and a sinusoidal wave shape. Positions 3 and 7 are the maximum values of induced e.m.f. in one side of the loop and it should be noted that polarities change at positions 1 and 5. The rotation therefore generates positive and negative values of electricity and it is this principle that is used in the construction of modern power station alternators.

Transformers

A transformer is a device for use on a.c. supplies. Its purpose is to transform energy from one voltage to another voltage. It operates on the principle of **electromagnetic induction**.

If you consider a coil connected to a d.c. source of supply, the current passing through it produces a magnetic field with a definite north and south pole at either end. But if the coil was connected to an a.c. supply the changing a.c. current would produce a changing magnetic field and this would have a changing north and south pole. By its own **self-inductance** a voltage would be created. If a second coil is placed next to the first coil, **flux linkages** would cut through it and a voltage would be induced in it by **mutual inductance**. The closer the coils are together, the greater will be the induced e.m.f. It can be made much greater by using a metal conducting path between the two coils in order to concentrate the magnetic flux linkages.

A laminated metal core and insulated coil windings are the essential parts in the construction of a transformer. The winding connected to the a.c. supply is called the **primary winding** and the one connected to the load is called the **secondary winding**. There is no electrical connection between these windings; they are in fact quite independent of each other. Figure 3.27 shows the magnetic field path through a double-wound transformer. The core is made of steel laminations to reduce heat losses caused by induced **eddy currents**. These are

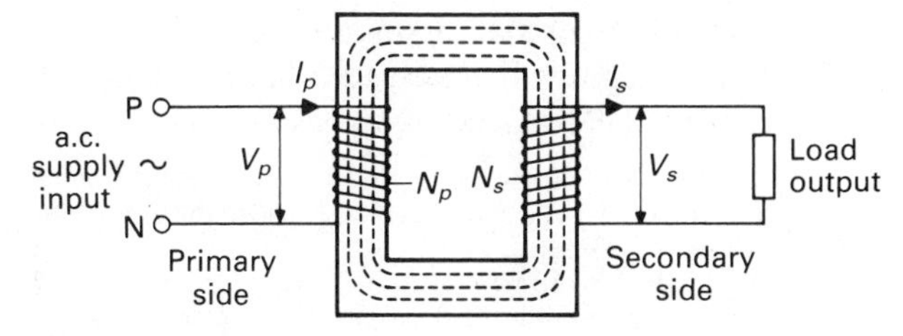

Figure 3.27 Double-wound transformer

circulating currents and the path they take is shown in Figure 3.28. Diagram (a) shows a solid magnetic core and diagram (b) shows a laminated magnetic core. If both cores had the same resistance, say 5 Ω and the circulating current in each were, say 5 A, the power loss in (a) would be

$$P = I^2R = 25 \times 5 = 125 \text{ W}$$

In (b), the circulating current in each lamination is only 1 A (since they are lightly insulated from each other). The loss in each lamination is

$$P = I^2R = 1 \text{ W}$$

and for the whole laminated core it is only 5 W. This is why it is necessary to keep all a.c. magnetic systems highly laminated.

Since the same magnetic flux links both cores, the induced e.m.f.s in each winding are proportional to the number of turns in each winding. In theory the no-load current of a transformer is very small, creating negligible voltage drop, and so the induced e.m.f. in both windings is almost equal to the voltages at the input and output terminals. The number of turns (N) and voltage (V) can be expressed as a ratio:

$$\frac{V_p}{V_s} = \frac{N_p}{N_s} \qquad [3.19]$$

where V_p is the primary voltage
V_s is the secondary voltage
N_p is the number of primary winding turns
N_s is the number of secondary winding turns

Formula [3.19] means that the voltage ratio is the same as turns ratio.

A transformer has no moving parts and is regarded as a highly efficient piece of static plant with an efficiency around 98%. Its chief heat losses are **iron losses** (caused in the transformer core) and **copper losses** (caused in the windings) and because of this its output power (P_O) is almost equal to its input power (P_I). Using formula [3.14] this can be written as

$$V_pI_p \cos\phi_p \approx V_sI_s \cos\phi_s$$

Since power factors ($\cos\phi$) are almost equal

$$V_pI_p = V_sI_s$$

Expressing this as a ratio

$$\frac{V_p}{V_s} = \frac{I_s}{I_p} \qquad [3.20]$$

This shows that the current ratio is inversely proportional to the turns ratio. If both [3.19] and [3.20] are combined, then the complete transformation ratio becomes

$$\frac{V_p}{V_s} = \frac{N_p}{N_s} = \frac{I_s}{I_p} \qquad [3.21]$$

In simple terms, this transformation ratio means a transformer with the same number of turns on each winding has the same primary to secondary voltages (i.e. it is a 1 : 1 transformer). When voltage is stepped up, the higher voltage winding will have more turns and current in this winding will be reduced. When the voltage is stepped down, there will be fewer turns in the lower voltage winding, but increased current.

One of the difficult things to understand about a transformer is what happens when it is connected to a load, i.e. when its secondary winding carries current. This secondary current creates another magnetic flux in the core, which tries to cancel out the flux produced by the current in the primary

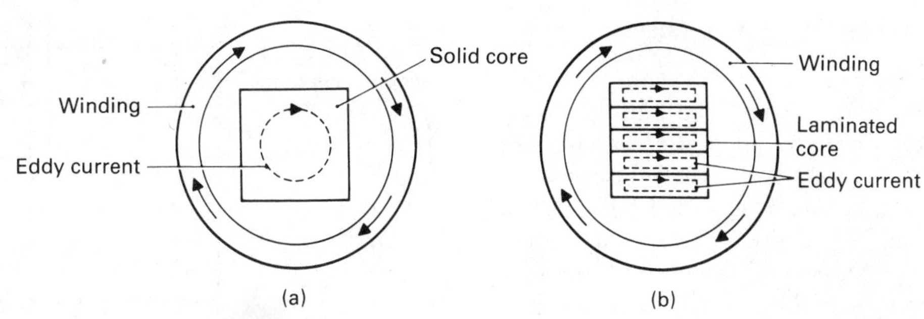

Figure 3.28 Paths taken by eddy currents

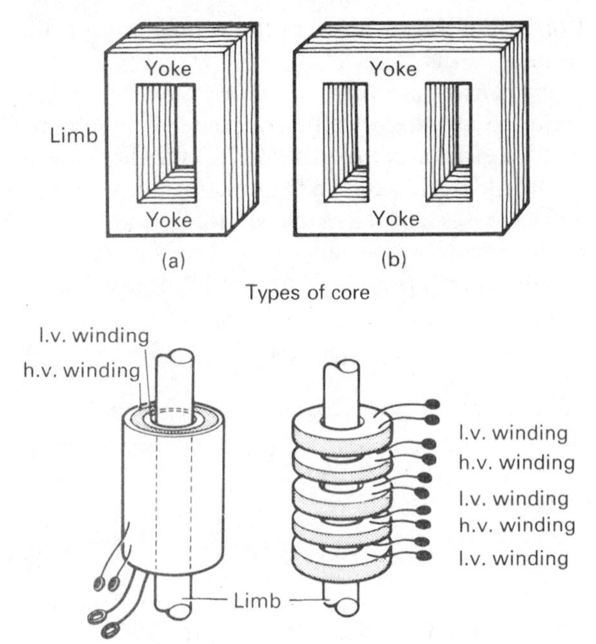

Figure 3.29 Transformer cores

winding. To restore the magnetic flux in the core the primary winding has to draw a current from the supply so that the e.m.f.s of both windings remain unchanged. This leaves the transformer with a **constant flux** at all loads and it also means that the transformer's iron losses will be constant at all loads.

Figure 3.29 shows the normal methods of designing double-wound transformers, (a) is a core type with two limbs and (b) is a shell type with three limbs, with the centre limb twice the size of the outer limbs. The windings are either of the concentric type or sandwich type as shown in (c). Attention should be paid to the low voltage and high voltage windings, since it is practice to place the low voltage winding next to the core ((c)(i)).

Chemical effects of electric current

The effect of an electric current flowing through a liquid is to try and separate the liquid into its chemical parts. This splitting up of the liquid's chemical compounds is a process called **electrolysis**. Faraday discovered this effect when he passed a current between two copper plates which were immersed in a copper sulphate solution. Any liquid solution like this, which can conduct electricity is referred to as an **electrolyte**. Faraday found that copper was deposited on the cathode plate (the negatively charged plate). From his experiments he deduced two laws:

1. The mass (m) of a substance deposited on an electrolytic cell is proportional to the current (I) and to the time (t) for which it passes.
2. The masses of different substances deposited or liberated by the same quantity of electricity (Q) are proportional to the various electro-chemical equivalents of the substances (z).

These two laws result in the formula

$$m = Itz \qquad [3.22]$$

Since $Q = It$ from formula [3.1]

$$m = Qz$$

By transposition

$$z = \frac{m}{Q} \qquad [3.23]$$

Figure 3.30 shows the arrangement of a **copper voltameter** used for determining the electro-chemical equivalents of metals. For copper it is found to be 0.3294 mg/C whereas for silver it is 1.1182 mg/C.

The process in which metals are deposited on the cathode electrode is called **electro-plating**. The idea is to cover a base metal, such as steel, with a very thin film of a much more expensive metal, which not only resists corrosion but also provides an attractive appearance, such as gold, silver, nickel and chromium platings.

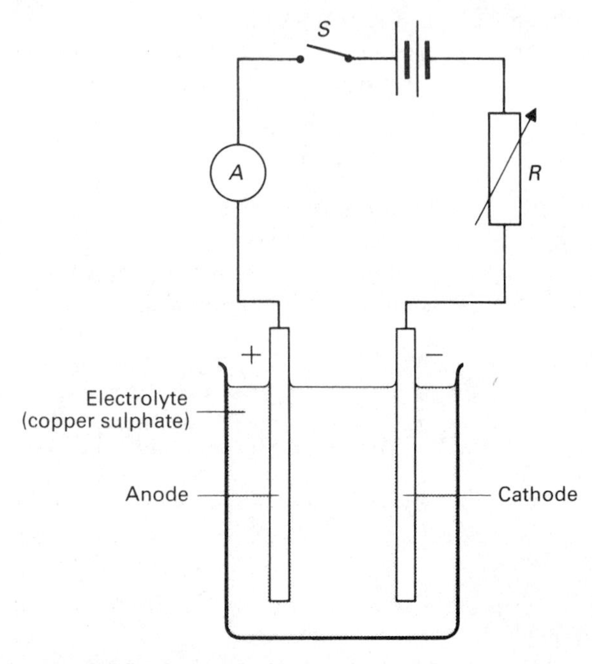

Figure 3.30 Copper voltameter

Secondary cells

A primary cell is a cell that has no further use when its chemical composition is exhausted, but a secondary cell is one that has the capability of being **recharged** by passing current through it. The two most common types of secondary cell are the **lead acid cell** and the **alkaline cell**.

Both these cells or batteries provide important commercial duties and apart from daily routine use such as in starting and running vehicles, they are frequently used as back-up for standby emergency lighting, as well as small power in such places as hospitals, cinemas, theatres and banks and many other important buildings.

The basic lead acid cell consists of two sets of plates immersed in an electrolyte. One type, known as a **Plante cell**, has a positive plate of pure lead and a negative lead plate which is a pasted grid containing lead oxides. The electrolyte used is **pure sulphuric acid**, which has a relative density of between 1.205 and 1.215 when the cell is fully charged. Figure 3.31 shows the construction of this cell, which has a nominal voltage of 2 V per cell. The final discharge voltage should not be allowed to fall below 1.8 V.

There are several other types of lead acid cell, such as the **flat plate** and **tubular plate**. The differences between them are in their positive plate arrangement, the type of container and their expected life cycle. For example, Plante cells have a life cycle of about 20 years in which they provide 100% capacity. They are somewhat bulky for a given ampere hour capacity and they are expensive. The tubular type is mechanically robust and less costly, but has a shorter life cycle of between 10 and 12 years. The flat plate type is also less expensive than the Plante type and it is more compact, but like the tubular plate type it suffers from having a shorter life cycle, of about 10 years. These three types of lead acid cell are all vented and designed to

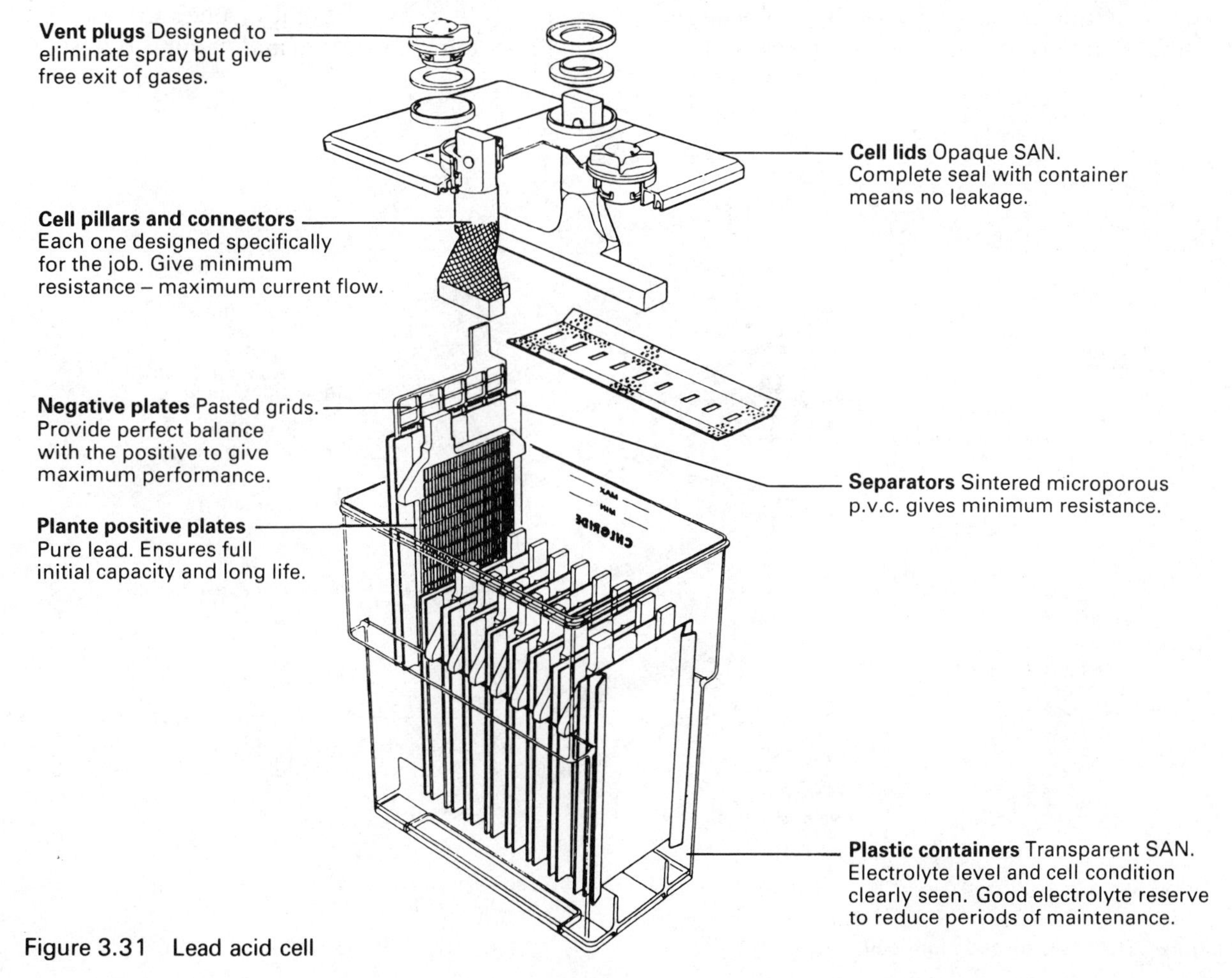

Figure 3.31 Lead acid cell

a capacity of several hundred ampere hours. One modern form of lead acid cell is the **sealed** type, which is used where there must be no risk of the electrolyte being spilled. They can be charged and discharged or stored in any position and whilst they are relatively cheap there are no maintenance requirements. Unfortunately, their life cycle is only four to seven years and their ampere hour capacity is not as high as the other types mentioned. They also lose this capacity with age and storage.

Alkaline cells are either **nickel–iron** or **nickel–cadmium**. They can be sealed or vented types. The sealed types have the advantage that they are gas tight and require no maintenance. They have an indefinite shelf life even when discharged. However, they are relatively expensive and they have a low ampere hour capacity (10 Ah) with a life expectancy of about four to seven years. The vented types have a much greater life expectancy of between 20 and 25 years and provide a good performance over a wide temperature range. They can also be left in a discharged state without being damaged but again they are a little expensive in comparison with the cost of Plante cells. They unfortunately suffer a wide voltage drop from 1.4 V per cell down to 1.0 V per cell.

One type of nickel–cadmium cell is the **Alcad cell**, which uses nickel hydrate combined with pure graphite for its positive plate. Its negative plate consists of cadmium oxide combined with a special oxide of iron. The electrolyte is pure **potassium hydroxide** in distilled water and it conducts current between the plates. Only distilled water should be used for topping-up purposes. The density of this cell's electrolyte is between 1.15 and 1.2 depending on the type and should not fall below 1.145. The nominal voltage per cell is 1.2 V. Smaller cells like the **Cyclon** type, which are rechargeable, provide 2 V per cell. They are sealed batteries with small ampere hour capacities.

CHARGING AND MAINTENANCE

There are several different methods of charging secondary cells. They are often divided into **constant current** and **constant potential** methods. The constant current method may be either slow charge or fast charge, but it involves the use of a charging source capable of maintaining a constant

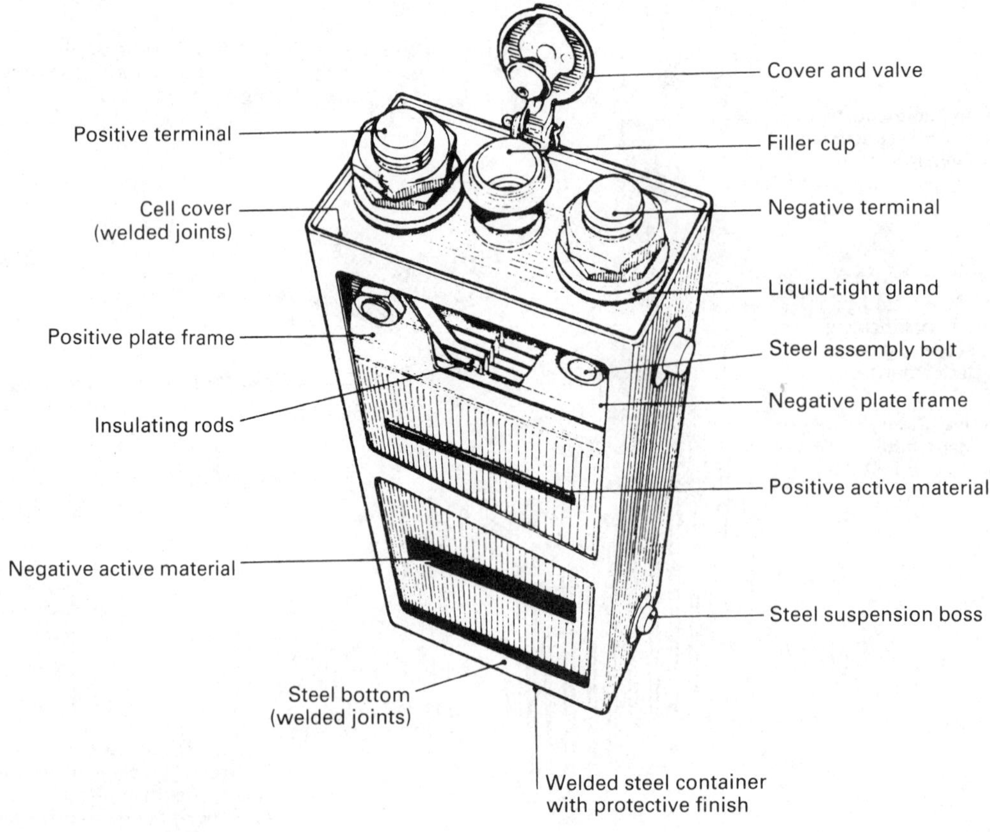

Figure 3.32 Nickel–cadmium cell

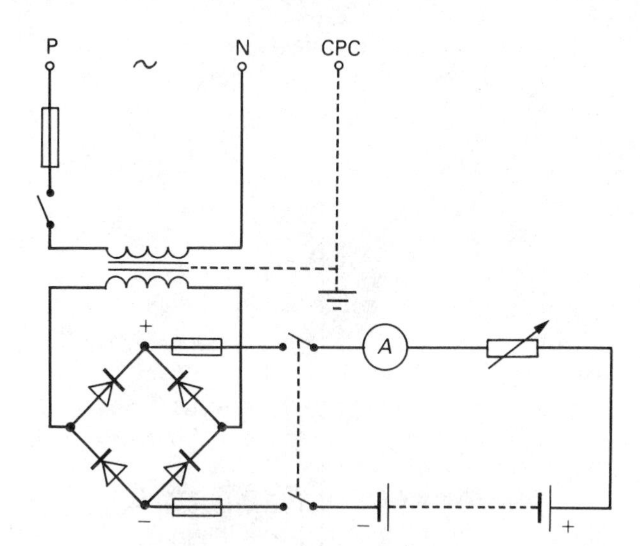

Figure 3.33 Battery-charging circuit

current throughout the charge period. The constant potential method of charging involves the use of a constant voltage source to the cells and the value of charge current will vary according to the state of the cells. The high performance Plante cells use either method: the slow charge is sometimes referred to as **trickle charging**, i.e. a method of keeping the cells fully charged by passing a small current through them that will not cause gassing or allow the density of the electrolyte to fall over a period of time. The constant voltage method is often referred to as **float charging**, i.e. keeping the voltage applied to the cells at 2.25 V per cell, and is the method used where continuous and variable loads exist. This method of charging has the advantage that it does not require manual attention. Figure 3.33 shows a circuit diagram of a simple battery-charging circuit using the constant current method. The series variable resistor, called a rheostat, is adjusted to increase or decrease the charge current through the battery. It is important to use a direct current source such as a rectifier unit and to make sure that its positive terminal is connected to the battery's positive terminal. A voltmeter can be connected across the battery to check its terminal voltage.

The following examples serve to illustrate the two methods of battery charging. Internal resistance has been ignored.

Example 3.16

Constant current method

Three 2 V lead acid cells were recharged from a 12 V d.c. supply source at a current of 4 A. When tested at the beginning of charge the terminal voltage of each cell was 1.9 V and at the end of the charging period it was 2.7 V. What was the value of the rheostat at the beginning and end of charge?

Solution

At the beginning of charge the cells' voltage

$$E = 3 \times 1.9 = 5.7 \text{ V}$$

The supply voltage to the cells, $V_s = 12$ V

Since

$$V_s = E + IR$$

then $$V_s - E = IR$$

$$12 - 5.7 = 4R$$

thus $$R = \frac{6.3}{4} = 1.575\ \Omega$$

At the end of charge the cells' voltage,

$$E = 3 \times 2.7 = 8.1 \text{ V}$$

and since

$$V_s - E = IR$$

$$12 - 8.1 = 4R$$

so $$R = \frac{3.9}{4} = 0.975\ \Omega$$

Example 3.17

Constant voltage method

Four 2 V lead acid cells are recharged from a 12 V d.c. supply source. At the beginning of charge, each cell voltage is found to be 1.9 V and the charging current is 4 A. What is the final current at the end of charge if each cell voltage has increased to 2.7 V?

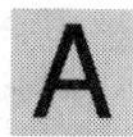

Solution

The constant voltage supplying the cells does not alter but the cells' voltage increases in opposition to it, and in doing so limits the current flowing.

At the beginning of charge the cells' voltage

$$E = 4 \times 1.9 = 7.6 \text{ V}$$

The supply voltage to the cells,

$$V_s = 12 \text{ V}$$

Since

$$V_s - E = IR$$

$$12 - 7.6 = 4R$$

$$R = \frac{4.4}{4} = 1.1\ \Omega$$

This resistance is a fixed value in the charger unit. At the end of charge the cells' voltage,

$$E = 4 \times 2.7 = 10.8 \text{ V}$$

Since

$$V_s - E = IR$$

$$12 - 10.8 = I1.1$$

and $$I = \frac{1.2}{1.1} = 1.09 \text{ A}$$

The state of charge of a lead acid cell is directly proportional to the density of its electrolyte (i.e. the ratio of its relative density to that of water) and this can be found by using a hydrometer, shown in Figure 3.34. It consists of a weighted bulb with a slender graduated stem and it floats vertically in the electrolyte being tested. The apparatus exposes a greater length of stem in an electrolyte with a higher density than in one with a lower density. In other words, the level at which it floats measures the density. In practice the electrolyte of a lead acid cell does not deteriorate throughout its life, but the unsealed nickel–cadmium types should have their electrolyte density checked every year since it does not vary with the state of charge, but falls gradually in service. Both lead acid and alcad types need occasional topping-up with distilled water and they give off hydrogen and oxygen during quick charge. You should not attempt to interchange electrolytes of these cells.

From the point of view of safety, you should keep a battery room well ventilated and the cells being charged should be kept clean and dry. Cell connections should be kept tight and covered with petroleum jelly in order to protect against corrosion. Cells should be recharged as soon as possible after discharge and records should be kept of each cell's voltage. You should avoid metal objects falling across terminals, as these could cause a spark, and

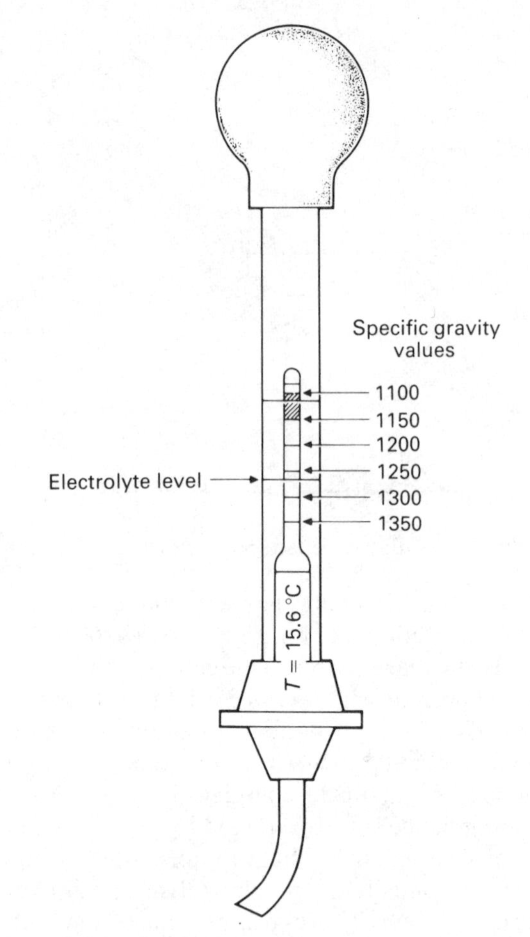

Figure 3.34 Cell hydrometer

no naked lights should be allowed in the room. When mixing or handling electrolyte, personnel should wear protective clothing such as goggles and rubber gloves.

INTERNAL RESISTANCE (r)

The internal resistance of a secondary cell is made up of the resistance of its plates, the connections and electrolyte. For large cells it is extremely low. For any type of cell it is found using the formula

$$r = \frac{(E - V)}{I} \qquad [3.24]$$

where E is the cell's e.m.f.
V is the cell's p.d.
I is the circuit current

It should be mentioned that a cell's p.d. is less than its e.m.f. when it is connected in circuit and this is due to its internal resistance. For example, the open circuit voltage of a lead acid cell is 2 V. When the cell is discharging at the rate of 10 A, its terminal voltage falls to 1.9 V. What is the cell's internal resistance?

Using [3.24]

$$r = \frac{(2 - 1.9)}{10} = 0.01\ \Omega$$

Figure 3.35 shows how cells can be connected to give either a series or parallel arrangement or a combined series–parallel arrangement. In the case of the series arrangement, all the cell e.m.f.s are added together and all the internal resistances are added together. If there are nine cells and each cell has an e.m.f. of 2 V and internal resistance 0.01 Ω, then the battery has an open circuit terminal voltage of 18 V and its internal resistance is 0.09 Ω. In the parallel connection the battery open circuit voltage is maintained at 2 V but its internal resistance is now only 1/9th of the original value, i.e. 0.0011 Ω. In the series–parallel arrangement, the battery voltage is 6 V and the internal resistance is now that of only one cell.

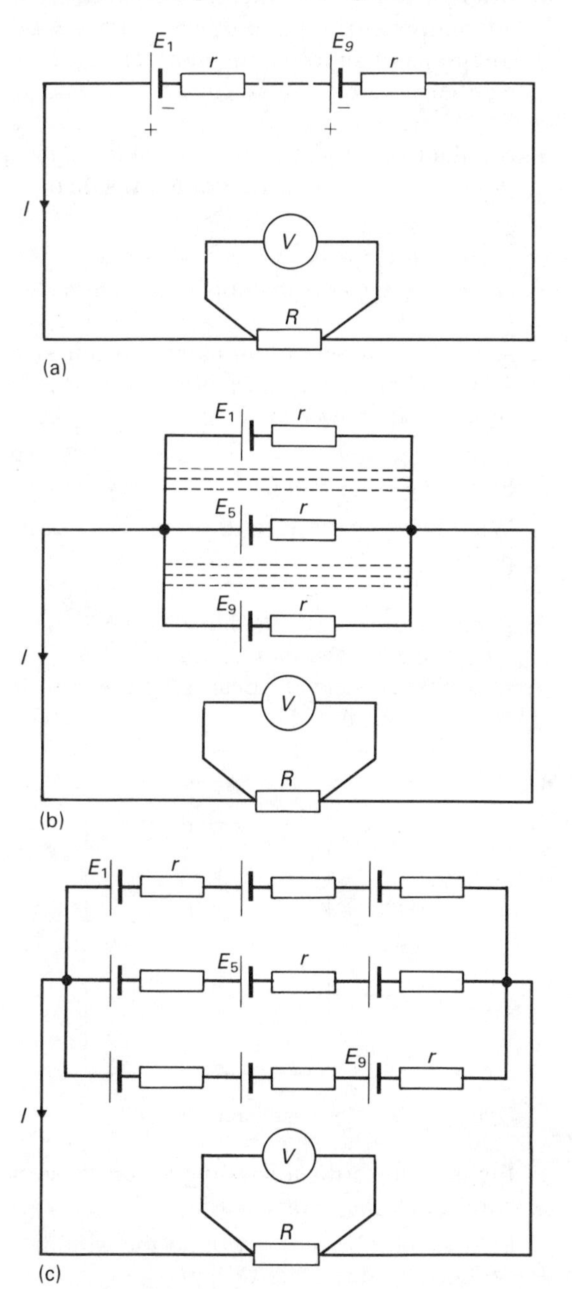

Figure 3.35 Cells connected in series, or series–parallel arrangements

EFFICIENCY

The **ampere hour capacity** is a term used to describe a cell's capability, but it really describes the ratio of the quantity of electricity taken out of the cell to the quantity of electricity required to be put back. It should be noted that a secondary cell is a **storage cell** and sometimes it is called a storage battery or accumulator. However, since **efficiency** usually expresses **output/input**, the cell's ampere hour (Ah) efficiency can be expressed as

$$\text{Ah efficiency} = \frac{\text{Ah given on discharge}}{\text{Ah required to charge}}$$

The Ah efficiency is always stated for a given set of discharge conditions. In practice, for most lead acid cells at the 10 hour rate of discharge it is about 90%. This means that an additional 10% needs to be put back into recharging the cell. One thing to note is that higher discharge currents give lower efficiencies.

Example 3.18

A lead acid cell is discharged at a rate of 3 A for 15 hours. If its efficiency is 90% what is the ampere hour charge required?

Solution

Discharge rate is $3\ \text{A} \times 15\ \text{h} = 45\ \text{Ah}$

Since

$$\text{Ah efficiency} = \frac{\text{Ah discharge}}{\text{Ah charge}}$$

then

$$\text{Ah charge} = \frac{\text{Ah discharge}}{\text{Ah efficiency}}$$

$$= \frac{45}{0.9}$$

$$= 50 \text{ Ah}$$

Another useful efficiency is called the **watt hour efficiency** and it expresses the ratio

$$\text{Wh efficiency} = \frac{\text{Watt hours during discharge}}{\text{Watt hours during charge}}$$

Example 3.19

A secondary battery is charged at a rate of 4 A for 10 hours and then discharged at 5 A for 6 hours. During charge the average battery voltage is 16 V while on discharge it is 12 V. Determine the battery's Ah efficiency and Wh efficiency.

Solution

Ah efficiency

$= \text{Ah discharge/Ah charge}$

$= (5 \times 6)/(4 \times 10)$

$= 0.75\ (75\%)$

Wh efficiency

$= \text{Wh discharge/Wh charge}$

$= (12 \times 5 \times 6)/(16 \times 4 \times 10)$

$= 0.56\ (56\%)$

Note: The Wh efficiency takes into consideration that the battery requires a higher voltage for charging.

EXERCISE 3

1. Three resistors of 8 Ω, 12 Ω and 24 Ω respectively, are connected across a 220 V supply. Determine the equivalent resistance when they are connected in a) series b) parallel. In each case find the power consumed.

2. A piece of copper wire 10 m long and 10 mm^2 in cross-sectional area carries a current of 5 A when it is connected to a 240 V supply. Determine the resistivity of the wire.

3. A coil has a resistance of 300 Ω at 0 °C. If its resistance is found to increase to 330 Ω at a temperature of 25 °C, determine its temperature coefficient of resistance.

4. Briefly explain the difference between alternating current and direct current. State what type of supply a transformer and rectifier requires.

5. A circuit has a total resistance of 0.06 Ω. When it carries a current of 60 A, calculate
   a) its voltage drop
   b) the power loss
   c) the energy consumed over 24 hours.

6. A 240 V fluorescent luminaire consumes 45 W. What current does it take from the supply if its power factor is
   a) 0.4 lagging?
   b) unity power factor?

   What device is used to improve the circuit power factor?

7. A motor drives a 3 kW hoist. If the efficiencies of the motor and hoist are 80% and 70% respectively, determine the electrical input to the motor.

8. 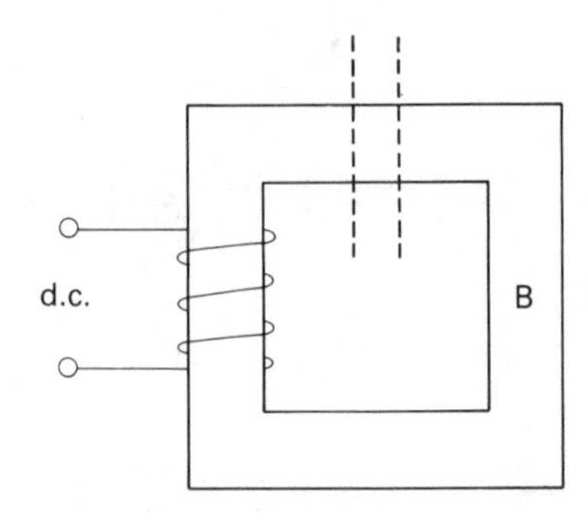

Figure 3.36 Magnetic circuit

Figure 3.36 shows a winding around a steel core. Copy the diagram and indicate

a) a current direction around the winding on each turn;
b) the direction of magnetic flux circulating the core.

What would be the effect of an equal and opposite flux produced by another similarly wound coil on the other limb of the core at B?

On either side of the dotted lines, indicate the conventional north and south pole as though the core were cut at this point.

9. A battery consisting of nine primary cells is connected to an external resistance of 10 Ω. If each cell has an e.m.f. of 1.5 V and an internal resistance of 0.45 Ω, determine the circuit current and voltage drop across the 10 Ω resistor when the cells are connected

   a) in series,
   b) in parallel and
   c) three sets in parallel, each set consisting of three cells in series (see Figure 3.35).

10. Draw a neatly labelled wiring diagram of a simple battery-charger circuit fed from a single phase a.c. supply, incorporating a double-wound transformer, bridge rectifier, ammeter and rheostat. Show also how a secondary battery is connected for charging purposes.

Basic electronics 4

Objectives

After reading this chapter you should be able to:

- *identify listed components used in electronic circuits.*
- *describe the method of indicating values of resistors and capacitors using colour and number codes.*
- *identify and describe the action of numerous types of semiconductor device.*
- *identify BS3939 electronic diagram symbols in common use.*
- *distinguish between alternating current and digital wave forms and state the relationship between frequency and periodic time.*

Electronic components

In the study of basic electronics, you have to recognise and identify the function of numerous circuit components such as resistors, inductors and capacitors. Some components are often mentioned in terms of their **active** or **passive** performance in a system. Active components within a circuit are those that have gain and are capable of directing current flow, for example, a number of semiconductor devices such as diodes, transistors and thyristors, and operational amplifiers found in integrated circuits. Passive or inactive components are basically resistors, capacitors and inductors.

Circuit devices used for converting one form of energy signal into another are called **transducers** and these are also separated into active and passive types. An active transducer is one that can generate some kind of pulse or signal, such as a photocell when it is exposed to light, whereas a passive transducer is one whose output response is capable of producing proportional changes, such as varying resistance, inductance or even capacitance. Transducers are at the heart of all electronic communication, but in this chapter only basic components will be discussed.

Resistors

A component whose chief property is resistance is called a **resistor** and is used to limit current flow in a circuit. Resistors for electronic work are small in terms of physical size, although they may have considerably high resistances. Selection for specific use is based on such factors as **ohmic value, power rating, stability** and **tolerance**.

Ohmic value is found by circuit design and application of Ohm's Law, which has been mentioned in Chapter 3. Power rating is the resistor's ability to perform in a circuit without overheating and is kept low to only a few watts. The resistor's stability concerns its ability to maintain the resistance value it had when it was originally designed. A lengthy service life, continual changes in working temperature and other environmental factors need to be considered in resistor design. Carbon is an ideal material since it is thermally stable and retains good mechanical strength at very high temperatures. Carbon also has a low expansion coefficient (see Chapter 2) and is a good conductor of heat. The tolerance factor is basically a product design allowance, taking into consideration a variation of the resistor's value slightly above or below its stated value. Different types of resistor have different tolerances, which range between 0.1% and 20%.

Resistors can be made to have either fixed resistance values (**fixed resistors**) or variable resistance values (**variable resistors**). Fixed resistors can be made from a carbon composition, carbon film or metal film, or be wirewound. Figure 4.1 shows some common types of resistor. Both carbon

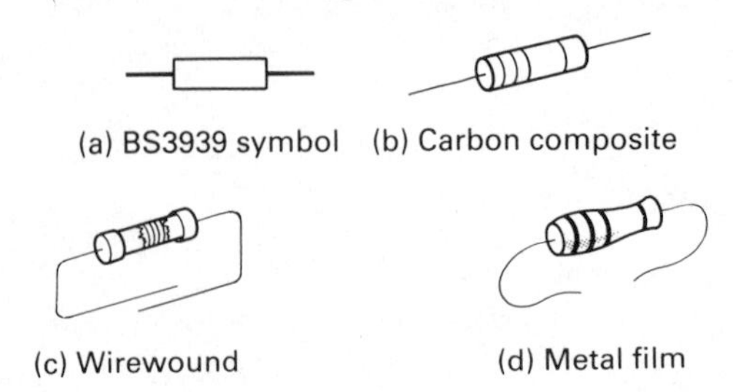

Figure 4.1 Types of resistor

Table 7 Resistor colour codes

Colour	*Numerical value*
black	0
brown	1
red	2
orange	3
yellow	4
green	5
blue	6
violet	7
grey	8
white	9
Tolerance band	
no band	±20%
silver	±10%
gold	± 5%
red	± 2%
brown	± 1%

types have general use, although the carbon film type has a much better power rating and greater stability. Its rugged construction makes it insensible to mechanical stress and environmental influences. Metal film resistors also have a wide general use, especially for specialised precision work such as instrumentation, where they need to be designed with much higher stabilities, low temperature coefficients and close tolerances.

Wirewound resistors are wound on ceramic tubes, which are often coated with silicon. Some are specially wound and sealed in epoxy resin to minimise the effects of inductance.

In practice, there are two methods of finding resistor ohmic values: one is by **colour code** and the other by **printed code**. The vast majority of carbon and metal film resistors in use today are marked with colour bands to indicate their resistance and tolerance values. Figure 4.2 and Table 7 explain this system, but reference should be made to Chapter 1 and Table 5, which provide the metric prefixes that must be used, especially M (mega), k (kilo), m (milli), and μ (micro). The printed code (see pp. 4–5) does not use bands but instead uses letters and numbers. This method finds additional use in circuit diagrams and also as a means of identifying variable resistors. Examples 4.1 and 4.2 explain both types of coding.

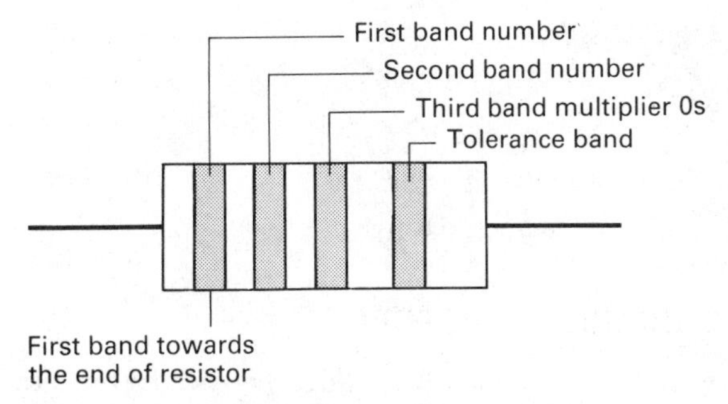

Figure 4.2 Resistor colour code

Example 4.1

What are the colour and printed codes for the following resistors?

a) 50 Ω + 1%; b) 220 Ω + 5%;
c) 3.9 kΩ + 10%; d) 10 MΩ + 20%.

Solution

Colour code

(Refer to Figure 4.2.) The first two bands are numbers, the third band represents the number of 0s and the last band is the tolerance. If there is no tolerance there won't be a band, but as a general guide to reading values, the first band colour will not be a tolerance colour such as silver or gold.

a) green, black, black and brown;
b) red, red, brown and gold;
c) orange, white, red and silver;
d) brown, black and blue.

Printed code

In this method the capital letter R denotes × 1, K denotes 10^3 and capital M denotes 10^6. The position of the letter gives the decimal point, and the last capital letter indicates the tolerance percentage (see Chapter 1).

a) 50RF, b) 220RJ, c) 3K9K,
d) 10MM.

Example 4.2

What are the values of the following resistors, marked

a) yellow, violet, orange and silver;
b) orange, white, blue and gold;
c) R33M;
d) 4M7K?

Solution

a) 47 kΩ + 10%;

b) 39 MΩ + 5%;

c) 0.33 Ω + 20%;

d) 4.7 MΩ + 10%

VARIABLE RESISTORS

Variable resistors, often called **potentiometers**, are designed for power ratings up to around 5 watts. Their main function in a circuit or system is to regulate the amount of voltage required. By adjusting a variable resistor you either increase or decrease the amount of current flowing in the circuit. A small resistance put in means a smaller voltage out; more resistance put in means more voltage out. Figure 4.3 shows a typical rotary spindle type variable resistor, which has a moving wiper in contact with a moulded carbon or ceramic-metal oxide (cermet) track. Some track designs are called **log**, in which the resistance change for equal angular clockwise rotations is greater at the end of the track than it is at the beginning. Others are called **linear** tracks, which exhibit equal changes of resistance as the carbon wiper rotates through equal angles. Some potentiometers are wirewound, while others are pre-set (often called **trimmers**), which are adjusted using a screwdriver. Trimmers are often designed with high stability and low rotational noise. Other factors such as heat, moisture and dust may also be considered.

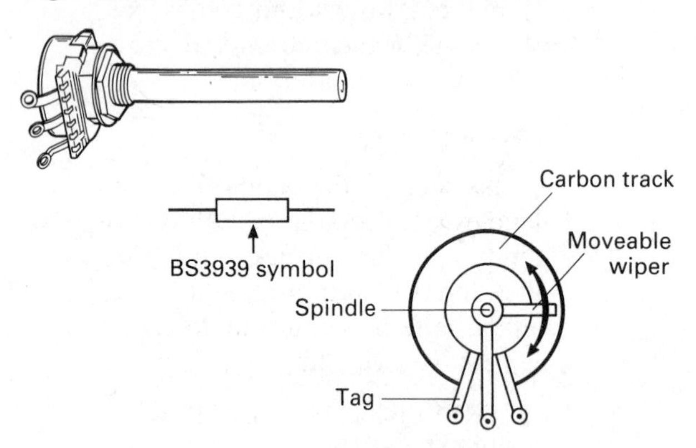

(a) Carbon track rotary type potentiometer

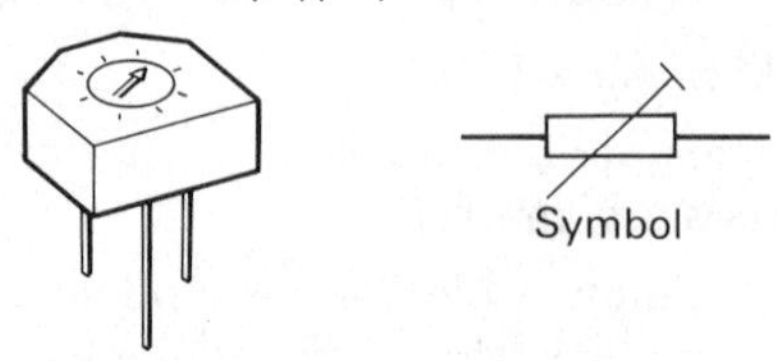

(b) Pre-set potentiometer trimmer

Figure 4.3 Types of variable resistor

Capacitors

A component whose chief property is capacitance is called a **capacitor** and is used to store electric charge. This can be of great benefit in electronic circuitry in, for example, controlling a.c. current, tuning radio receivers, time delay circuits and the separating of a.c. from d.c. Selection of a capacitor is based on its designated **working voltage** and its **leakage current**. Its working voltage is the maximum voltage it can take before its **dielectric** breaks down. The dielectric is the name given to the insulating medium between the plates of the capacitor and it allows opposite charges to be maintained on each of its plates. In Figure 4.4 the plate connected to the positive battery terminal will lose electrons and the plate connected to the negative battery terminal will gain electrons. During a short period of time, the voltage across the capacitor's plates will be the same as that of the battery and no further current will flow. The capacitor is said to be charged up – it stores electric energy.

When an alternating current is applied to the plates of a capacitor, there is a continual charge and discharge from each plate because the a.c. supply is cyclic in nature, changing its polarity each half cycle. The insulating dielectric between the plates is subject to an alternating electric stress. This results in heat dissipation and energy loss called **dielectric loss**. The leakage current factor is the current flowing between both plates, which results in loss of charge. This is normally very low in most low voltage capacitors and only a few microamperes in electrolytic capacitors. Capacitors should also have high resistivity, high insulation resistance and electric strength.

Like resistors, capacitors can also be designed to have fixed values or variable values. Fixed

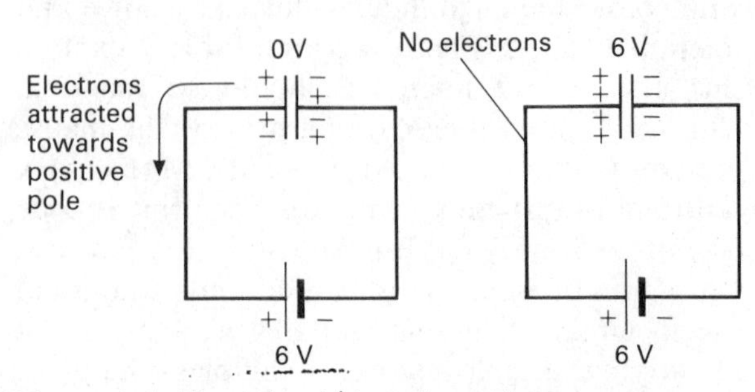

Figure 4.4 Capacitor being charged

capacitors can be divided into those which are **polarised** and those which are **non-polarised**. Polarised types include the standard **aluminium electrolytic** and **tantalum electrolytic** capacitors. The former are widely used in power supplies. They have separate positive and negative terminals and must be correctly connected in order to maintain their dielectric by electrolytic action.

In the aluminium type (see Figure 4.5) the dielectric is paper soaked in electrolyte of ammonium borate.

Non-polarised capacitors such as the **polypropylene, polycarbonate, polyester, polystyrene, mica** and **ceramic** types can be connected either way round. They all have extremely good dielectric properties with very small leakage. Ceramic and mica capacitors are not constructed like the polarised types. Silver mica types have silver deposited on the insulation and are used where high frequency and high stability is required, such as in tuned circuits and filter circuits. Ceramic capacitors have a metal surface deposited on the insulation and are suitable for use in high frequency, decoupling and suppression circuits. Figure 4.6 shows several different types available.

Capacitance values are either marked on the components or found by the colour code method. For polyester capacitors, the first two colour bands represent tens and units. The third band represents the multiplier expressed in picofarads (pF), the fourth band is the tolerance and the fifth band denotes the working voltage (see Figure 4.7 and Table 8). For a capacitor coloured orange, red, brown and white, the value would be 320 pF + 10%. For ceramic types, coloured dots are used. If the base of the capacitor is coloured red its working

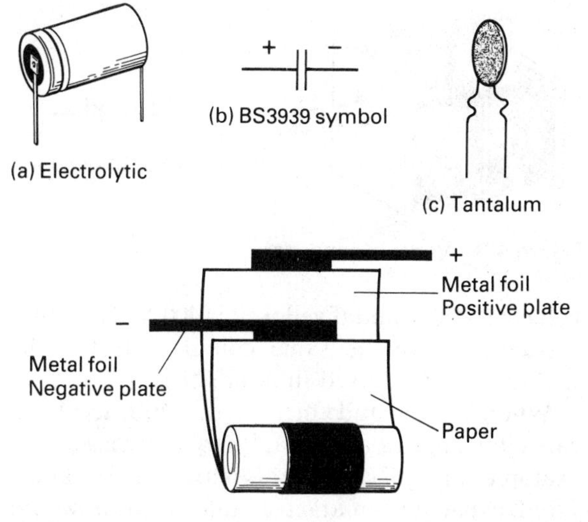

Figure 4.5 Polarised capacitors

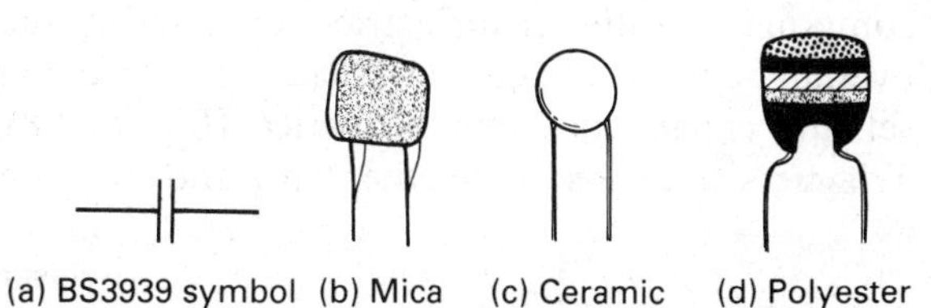

Figure 4.6 Non-polarised capacitors

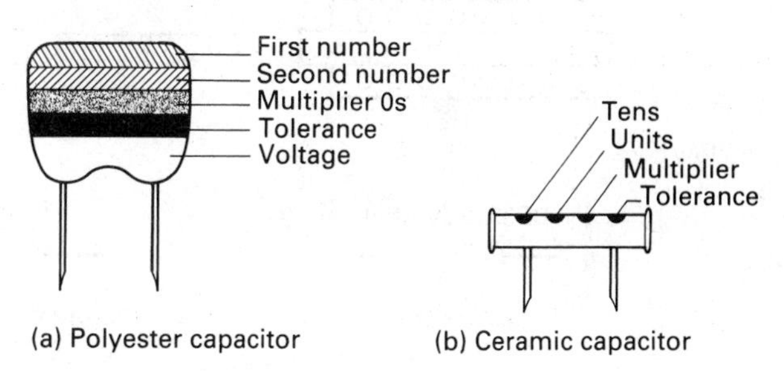

Figure 4.7 Capacitor colour codes

Table 8 Capacitor colour codes

Colour	*Numerical value*	*Multiplier*	*Tolerance*	*Voltage*
black	0	× 1 pF	±20%	
brown	1	× 10 pF	±1%	
red	2	× 100 pF	±2%	250 V
orange	3	× 1000 pF	±2.5%	
yellow	4	× 10 000 pF	—	400 V
green	5	× 1 000 000 pF	±5%	
blue	6		—	
violet	7		—	
grey	8		—	
white	9		±10%	

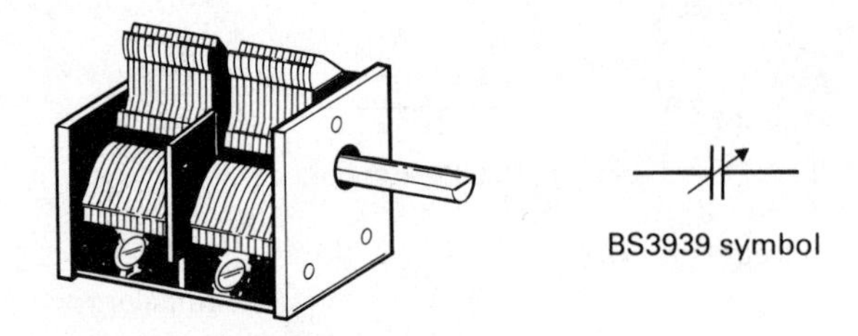

Figure 4.8 Variable capacitor

voltage is 250 V and if yellow it is 400 V. Tantalum capacitors have the same colour code but the multiplier is expressed in farads (F).

Where letter and digit codes are used on capacitors, p, n, μ and m are frequently used. For example, a 1.5 pF capacitor is marked as 1p5 and a 1.5 μF capacitor is marked as 1μ5 ; a 250 nF would be marked 250n.

Variable capacitors are used for tuning radio receivers and these are basically multi-plate components with air-dielectrics. By altering the overlap between a fixed set of plates and a moving set, the capacitance can be varied (Figure 4.8). **Trimmers** and **pre-sets** are other types and are often much smaller. Some trimmers are made with polypropylene dielectrics for high insulation resistance and low temperature coefficients, whereas pre-set components often use mica sheets compressed between metal foils.

Inductors

A component whose chief property is inductance is called an **inductor**. An inductor opposes changing currents through electromagnetic induction. It is used in electronic circuits for the production of high frequencies in, for example, radio tuners, and chokes, which are used to stop radio frequency currents. A transformer is a similar component, which also operates on electromagnetic induction, and is used for stepping up or stepping down voltage. In electronic work it is the main component in an audio frequency amplifier circuit and it is also used in radio frequency tuning circuits.

Selection of inductors and transformers is based on a number of factors according to their application. The essential parts of both these com-

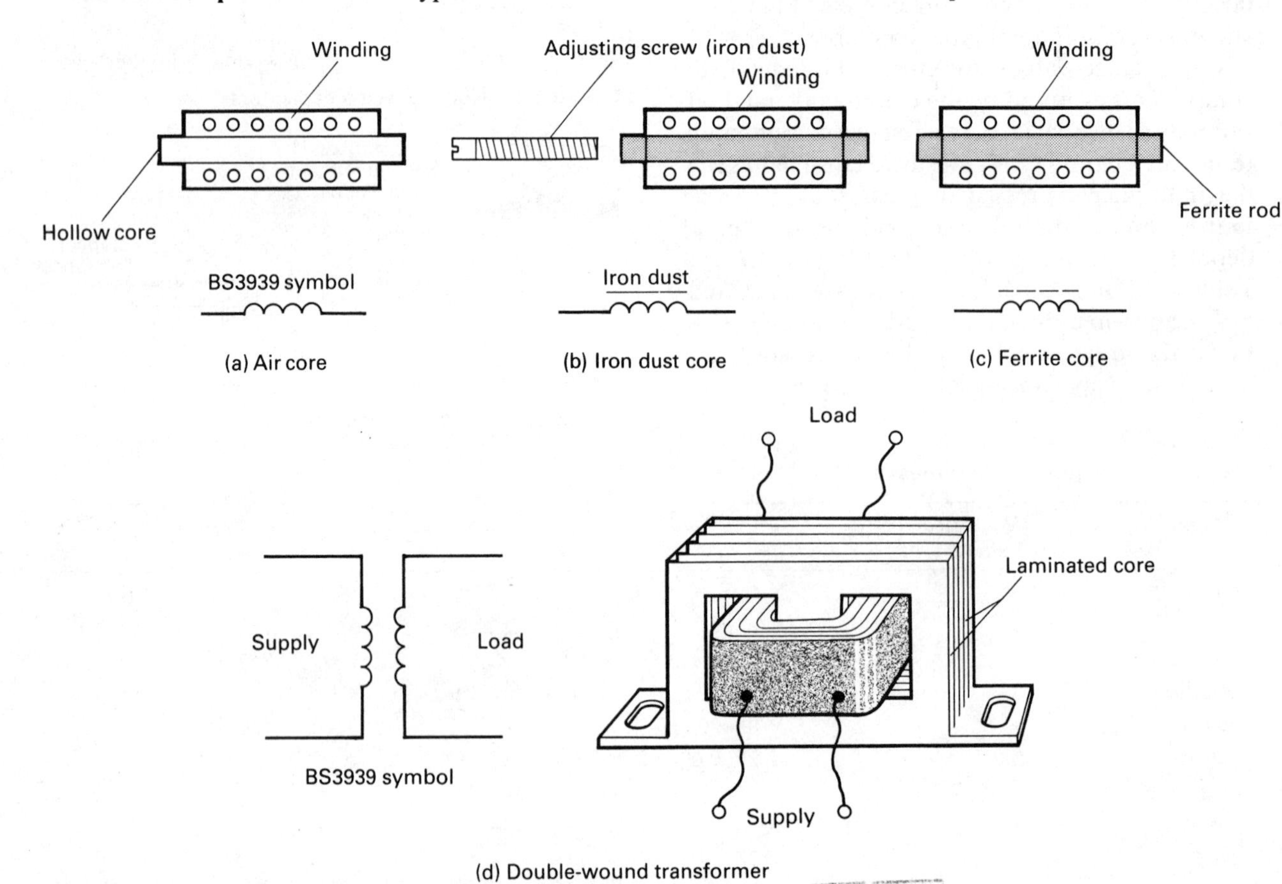

Figure 4.9 Inductors and transformers

ponents are the windings, which possess resistance and also reactance, which both oppose the flow of current in different ways. In a.c. circuits, current flow is affected by the combination of these two quantities, called **impedance**. In an inductor, the winding's **self-inductance** plays an important part by making the circuit current lag behind the supply voltage, creating a phase shift between supply current and supply voltage, which causes a low power factor (see Chapter 3).

Inductors may be **air-cored, iron-cored** or **ferrite-cored** (see Figure 4.9). Air-cored types have small inductances and their application is suitable for high frequencies. At high frequency an inductor's reactance becomes very large ($X_L = 2\pi fL$) and for this reason inductors are used as chokes to separate radio frequencies from audio frequencies. Iron-cored types produce a much greater magnetic field strength than air-cored types and consequently their inductances are much larger. They are suitable as chokes in low frequency smoothing circuits and in power supplies. Ferrite-cored inductors are constructed from non-conducting magnetic material and are also used in radio tuning circuits.

Transformers are components supplied with a.c., which operate on the principle of **mutual inductance**. They may have one or more windings. The one connected to the supply creates magnetic flux, which is used to induce a voltage in the second winding, which is connected to the load terminals. (For more on transformers, see Chapter 3). Some transformers only have one winding designed with tappings or a sliding contact in order to divide up the single winding to obtain a separate supply connection and load connection (see Figure 4.10). The a.c. is supplied through its **primary winding** and the load connections are taken from a section of the main winding called the **secondary**. A double-wound transformer has two distinct windings (called primary winding and secondary winding). These windings may also be designed with tappings and they can also be completely isolated from each other (isolating transformer). One important selection factor is their **voltampere rating** but physical size, core construction and class of insulation all need to be considered. Small double-wound transformers are used in mains circuits as well as providing matching in amplifier–loudspeaker circuits.

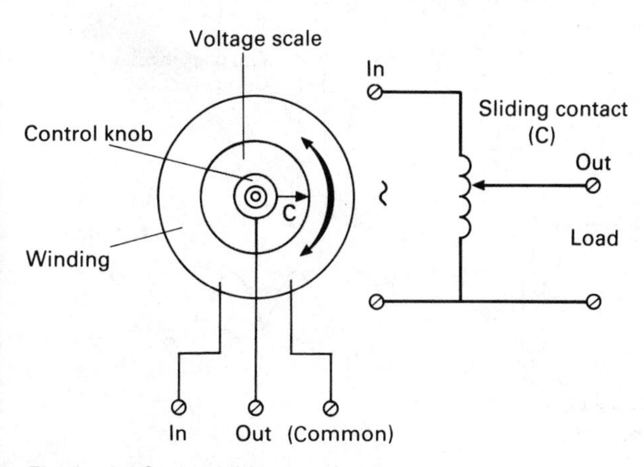

Figure 4.10 Variable transformer

Semiconductor devices

DIODE

This is a solid-state semiconducting device known in the electronics industry as a **p–n junction diode**. Diodes function either as open or closed switches on d.c. supplies, or as rectifying elements on a.c. supplies. The letters **p–n** represent the treatment process of the semiconductor material. Figure 4.11 shows these two materials, first apart and then together to form a junction. The p-type and n-type materials are formed by doping pure semiconductor material (such as silicon and germanium) with controlled amounts of impurity material (such as phosphorus and boron). Once formed, p-type material will contain surplus **holes** while n-type material will contain surplus **electrons**, but both materials are electrically neutral.

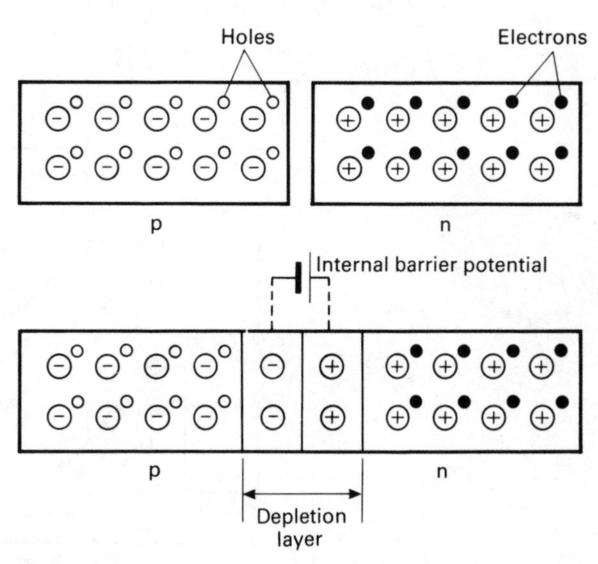

Figure 4.11 Formation of a semiconductor junction diode

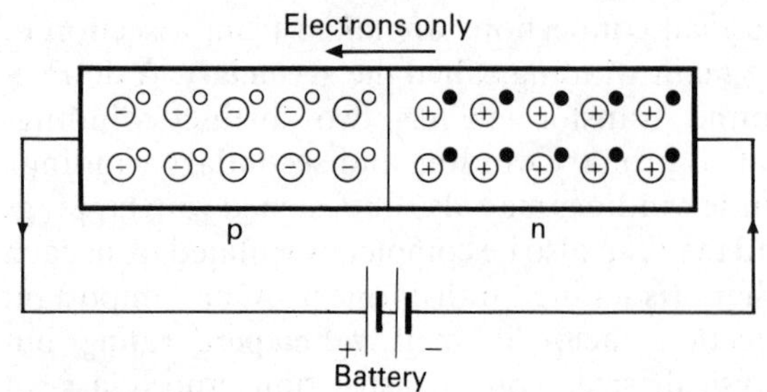

No internal barrier potential as a result of forward bias

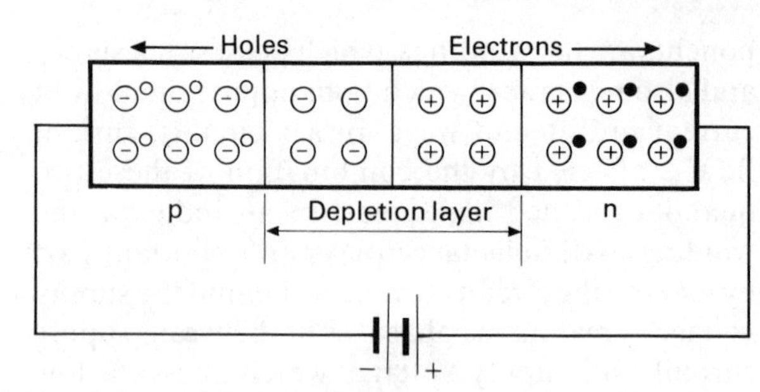

Increased depletion layer as a result of reverse bias

Figure 4.12 Diode connections

When the two materials are brought together to form a junction (**junction diode**), a small **barrier voltage** is created, similar to that of a small cell with a p.d. of only a few tenths of a volt. This barrier voltage is the result of unlike charges congregating at the junction. To break this down for conduction a silicon diode requires about 0.6 V, whereas a germanium diode requires about 0.3 V. Figure 4.12 shows the connections of the diode to a d.c. source. Depending on the connections made, it will either conduct current or stop current flow. These two modes of connection are called **forward bias** and **reverse bias**. The former allows current to flow one way (holes and electrons cross over the junction,

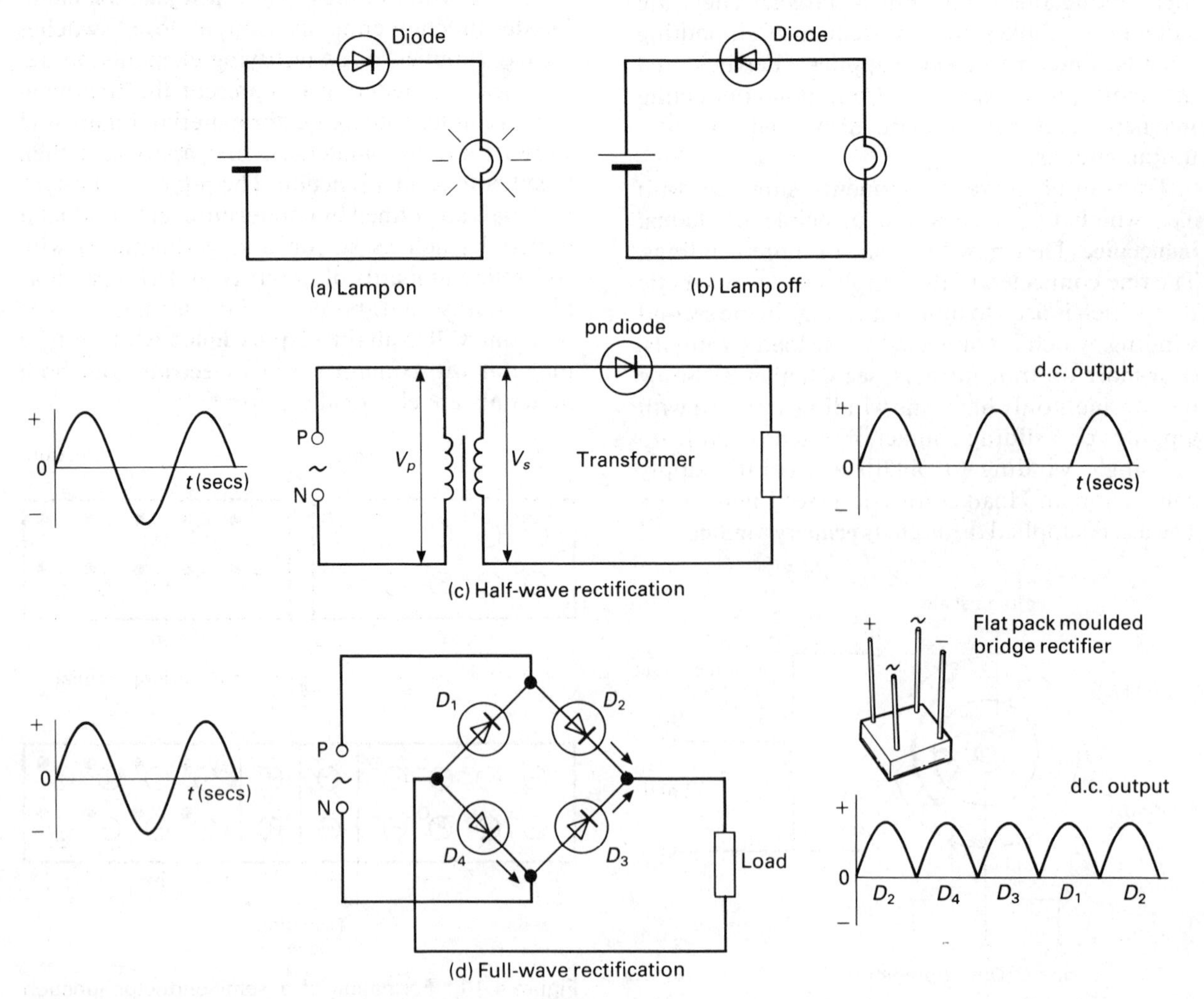

Figure 4.13 Diode connections on a.c. and d.c. supplies

breaking down the internal barrier), but when its connections are reversed or the d.c. supply connections are reversed, the holes in the p-region are attracted towards the negative terminal of the battery, and the free electrons in the n-region are attracted towards the positive terminal of the battery. This leaves a wide depletion layer in which current cannot flow. If the supply is a.c., only a positive half-pulse passes through the diode, which acts as a half-wave rectifier. Figure 4.13 shows the connections of a single diode acting as a switch in a d.c. circuit and as a half-wave rectifier element in an a.c. circuit. Its structure and characteristics are shown in Figure 4.14. The diode's cathode (shown on the p-material) is often marked with a band or identified by the design of its enclosure (arrow-shaped to show conventional current flow). It should be noted that silicon diodes are often preferred to germanium diodes in rectifier circuits because of their higher working temperatures and higher breakdown voltages. In situations where high current diodes are used as rectifier elements, it is recommended to use heat sinks which dissipate heat while they are on load.

ZENER DIODE

This device is similar to the ordinary diode but is used to control or maintain a constant voltage across loads. It is often called a **voltage regulator** or **stabiliser**, since its function is to maintain the output voltage at a fixed value. It is normally used in its reverse mode and designed to operate at its reference voltage V_z (the voltage at which the zener diode begins to conduct in its reverse mode). A current-limiting resistor is connected in series with it to ensure that its power rating is not exceeded; otherwise it will be damaged.

Figure 4.15 shows the zener diode in a typical circuit. If the zener diode has a reference voltage of 10 V and the battery voltage is 6 V, no current passes through the zener and V_z is also 6 V. If the battery voltage is increased above 10 V (say 12 V) then the zener diode conducts and V_z becomes

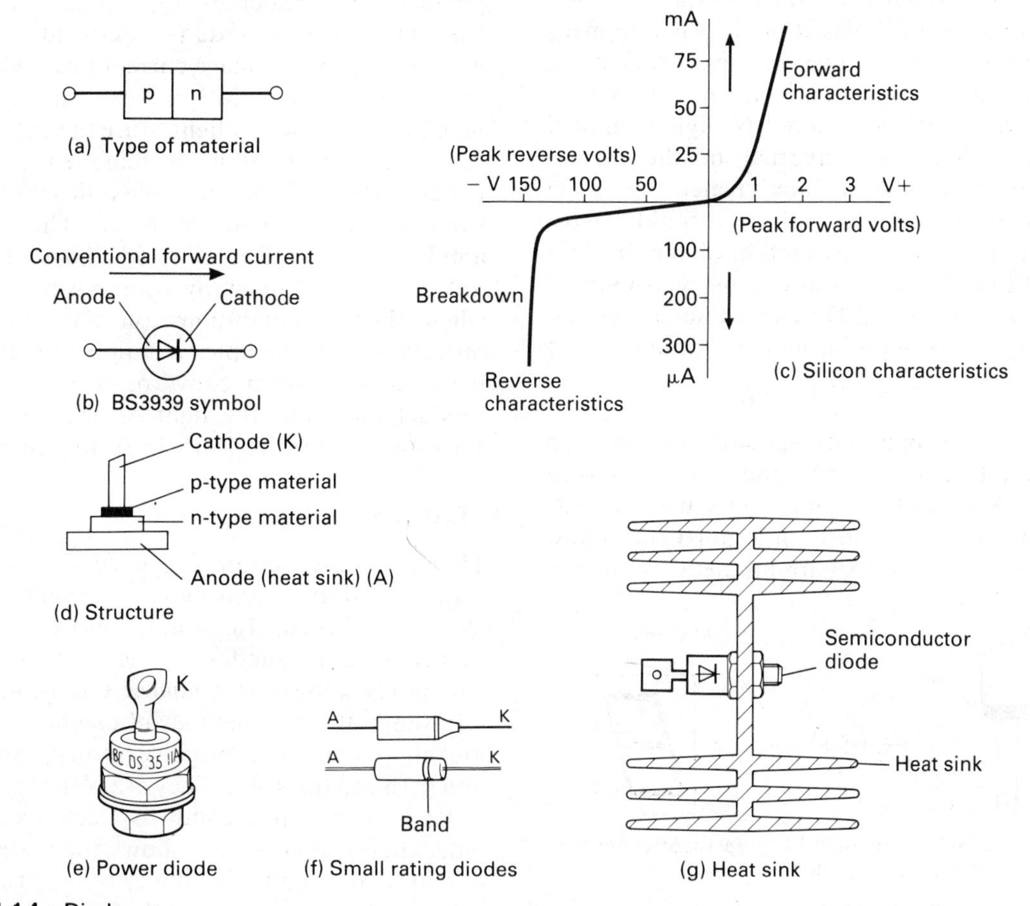

Figure 4.14 Diode

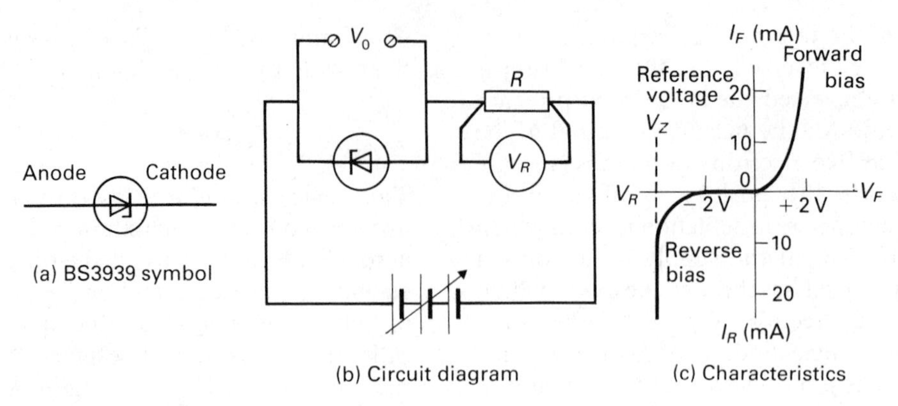

Figure 4.15 Zener diode

10 V. The limiting resistor (R) then has a p.d. across it of 2 V. If the battery voltage is increased to 14 V, then V_z will still remain at 10 V and the p.d. across R will increase to 4 V.

LIGHT EMITTING DIODES

These devices, commonly known as **LEDs**, have a coloured translucent enclosure which emits either red, yellow or green light when conduction takes place in the forward-biased mode. This happens when electrons release their energy as they move across the diode's junction in the direction from the n-material to the p-material. No light is emitted when the diode is connected in the reverse direction. LED colours are created by using gallium phosphide and gallium arsenide phosphide, and are frequently used in digital displays and indicator lamps. A resistor must be connected in series with the LED. The value of resistor necessary can be found from the formula

$$R = (V_s - V_f)/I_f$$

where V_s is the supply voltage, V_f is the voltage drop across the LED (about 2 V) and I_f is the forward current. Red LEDs require forward currents between 5 mA and 25 mA, while green and yellow LEDs require forward currents between 10 mA and 40 mA. Figure 4.16 shows a typical miniature LED suitable as a signal indicator on a printed circuit board or electrical accessory. Another use of LEDs is in displays, and Figure 4.16 also shows the LED segment appearance for numbers.

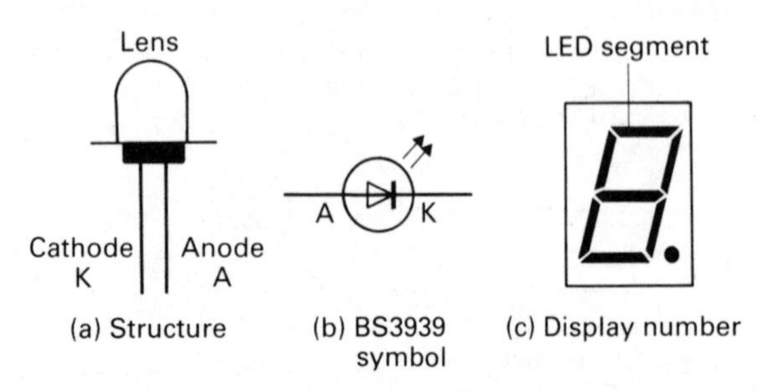

Figure 4.16 Light emitting diode (LED)

PHOTODIODE

This is another junction diode, which utilises photoelectric properties to generate a signal. Normally, when a diode is connected in reverse bias only a small leakage current flows. However, this small current will increase if the diode junction is exposed to light, since the light energy causes electrons to fill holes in the semiconductor material. Photodiodes are made with a transparent window for the light to enter. They require amplification to increase the signal output. Photodiodes are used in many counting type circuits where there is continuous 'on' and 'off' switch operation; for example, a number of products being counted on a conveyor belt, where each product interrupts the light on the photodiode. Figure 4.17 shows the device in an amplifier circuit.

TRANSISTOR

This is a three-terminal semiconductor device widely used in electronics as an amplifier or electronic switch. In simple terms, it is two semiconductor diodes joined back-to-back, commonly known as a **bipolar transistor**. Figure 4.18 shows the two methods of combining p-type and n-type material to produce (a) an **npn transistor** and (b) a **pnp transistor**. They are both very similar – the former mainly conducts electrons and the latter, holes. Figure 4.19 shows their input and output connections in three types of circuit: **common base, common collector** and **common**

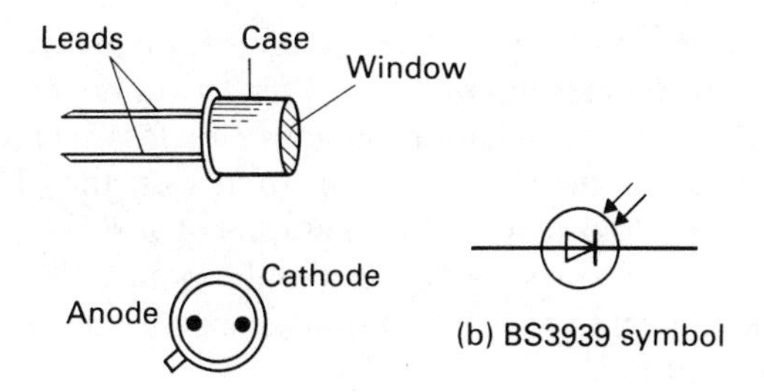

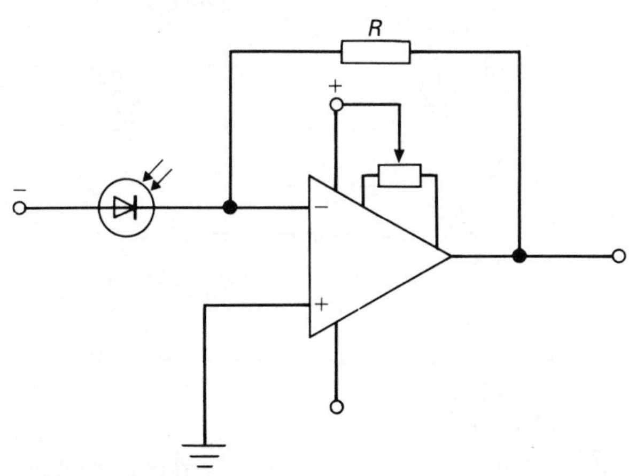

Figure 4.17 Photodiode

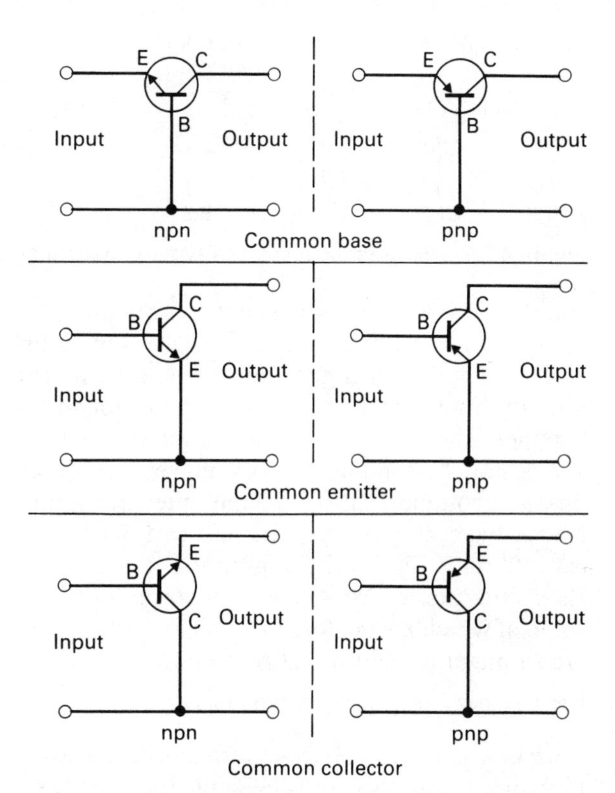

Figure 4.19 Methods of connecting bipolar transistors

emitter. The arrows indicate conventional current flow, not electron flow. The three connection leads are **base** (B), **collector** (C) and **emitter** (E).

In terms of operation, consider the following. Figure 4.20 shows the circuit connections of an npn common base transistor with its base connected to a 0.36 W/6 V lamp in series with a 1 kΩ resistor. In the collector circuit is another lamp of similar rating. With the common base and emitter terminals connected to a d.c. battery of 6 V, it is found that only lamp L_1 lights. If lamp L_2 is unscrewed to cause a break in the base circuit, L_1 goes out. When L_2 is inserted again, L_1 lights up as before. The unscrewed lamp acts as a switch. Only a very small current (say 1 mA) in the base circuit is sufficient to cause a much larger current (about 60 mA) to flow in the collector circuit. In this way the transistor acts as a current amplifier.

Any kind of triggering arrangement can replace lamp L_2 to switch on the transistor automatically. To give warning or signal of something happening, lamp L_1 could be replaced by a bell. There are many applications where this part of the circuit

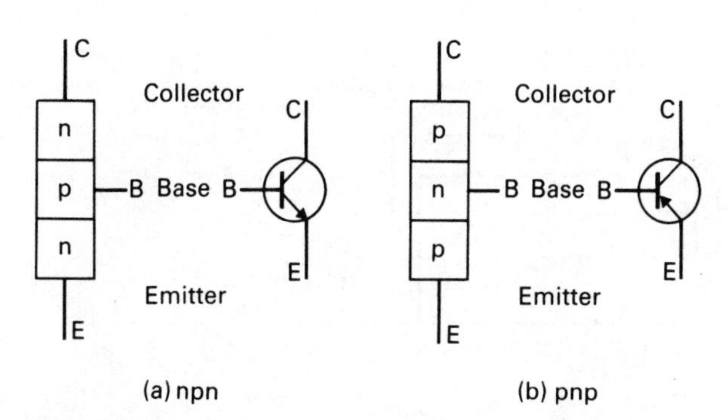

Figure 4.18 Types of transistor

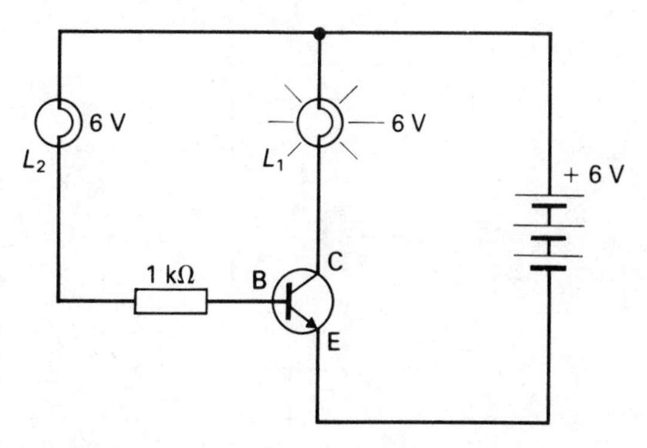

Figure 4.20 Circuit connections for a common base npn transistor

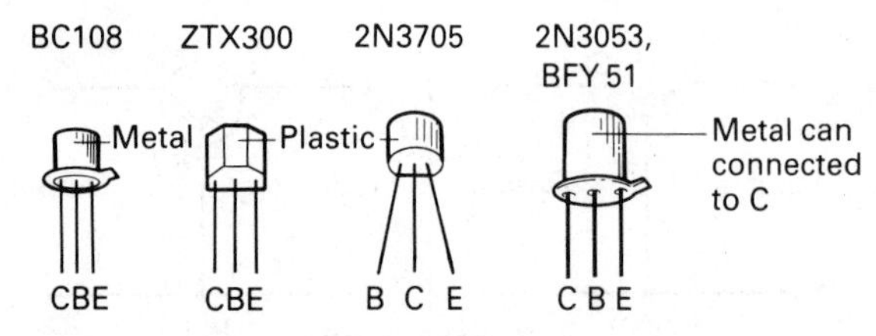

Figure 4.21 Pin lettering for several types of transistor

could be switched 'on', for example, the bending of contacts in a heat-operated switch, light shining on a photocell, the closing of contacts in a fire alarm unit or time switch, or float contacts touching together when a certain liquid level is reached. Transistors have numerous advantages over other physical communication systems: they are small, cheap, have no moving parts and can operate at exceptionally fast switching times. Figure 4.21 shows the pin lettering of several common types, some of which are enclosed in plastic while others are in metal for additional protection.

FIELD EFFECT TRANSISTOR (FET)

This is a **unipolar** voltage-controlled device with high input resistance and negligible input current, that is, the input voltage controls the output current. Figure 4.22 shows the basic construction of this component, which consists of a channel of n-type semiconductor material with two connections, source (S) and drain (D). A third connection is made at the gate (G), which is made of p-type material to control the n-channel current.

In theory, the drain connection is made positive with respect to the source and electrons are attracted towards the D terminal. If the gate is made negative there will be reverse bias between G and S, which will limit the number of electrons passing from S to D. The gate and source are connected to a variable voltage supply, such as a potentiometer, and increasing or lowering the voltage makes G more negative or less negative, which in turn reduces or increases the drain current. Figure 4.23 shows graphs of various gate-source voltages and you should notice that the more negative is V_{GS}, the less will be the drain current I_D.

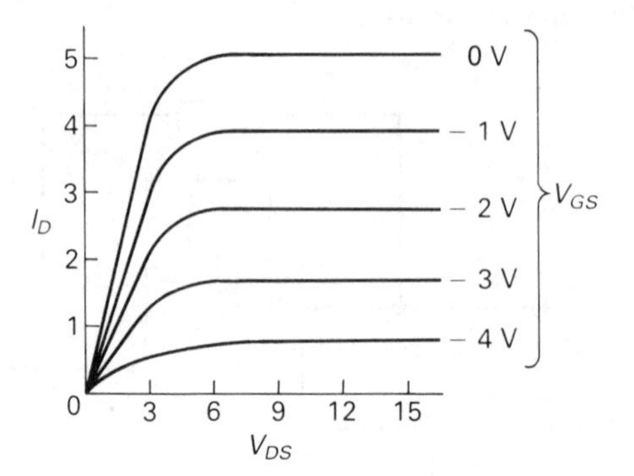

Figure 4.23 Gate–source voltages of an FET

THYRISTOR

This is a four-layer semiconductor device that is used as an electronic switch on d.c. supplies or a controlled half-wave rectifier on a.c. supplies. As a half-wave rectifier element made from silicon, it is referred to as a **silicon controlled rectifier** (SCR).

A thyristor behaves like a diode when conducting in its forward direction, but it will not conduct until a positive voltage is applied to its gate terminal (G). Once it has been triggered, the gate voltage can be removed and the thyristor will continue to conduct until switched off by the supply being disconnected or reversed. Figure 4.24 shows its structure, symbol and appearance as a high current device in rectifier circuits, and low current device in control circuits. The waveform shows how the output current can

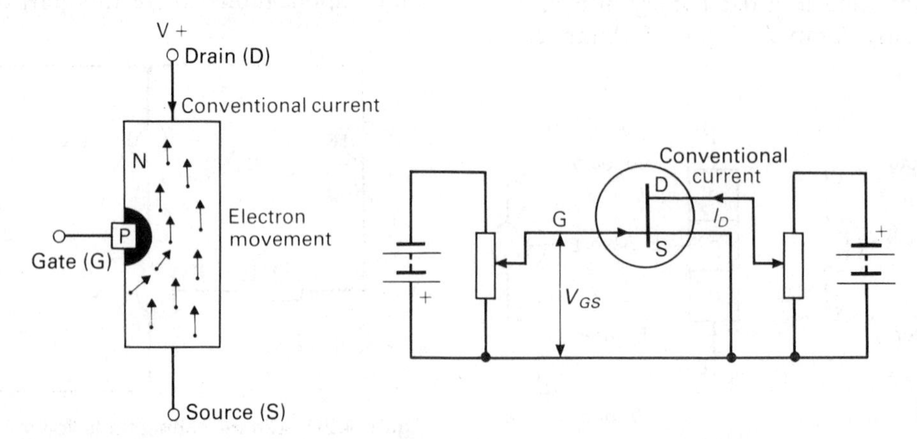

Figure 4.22 FET structure and circuit

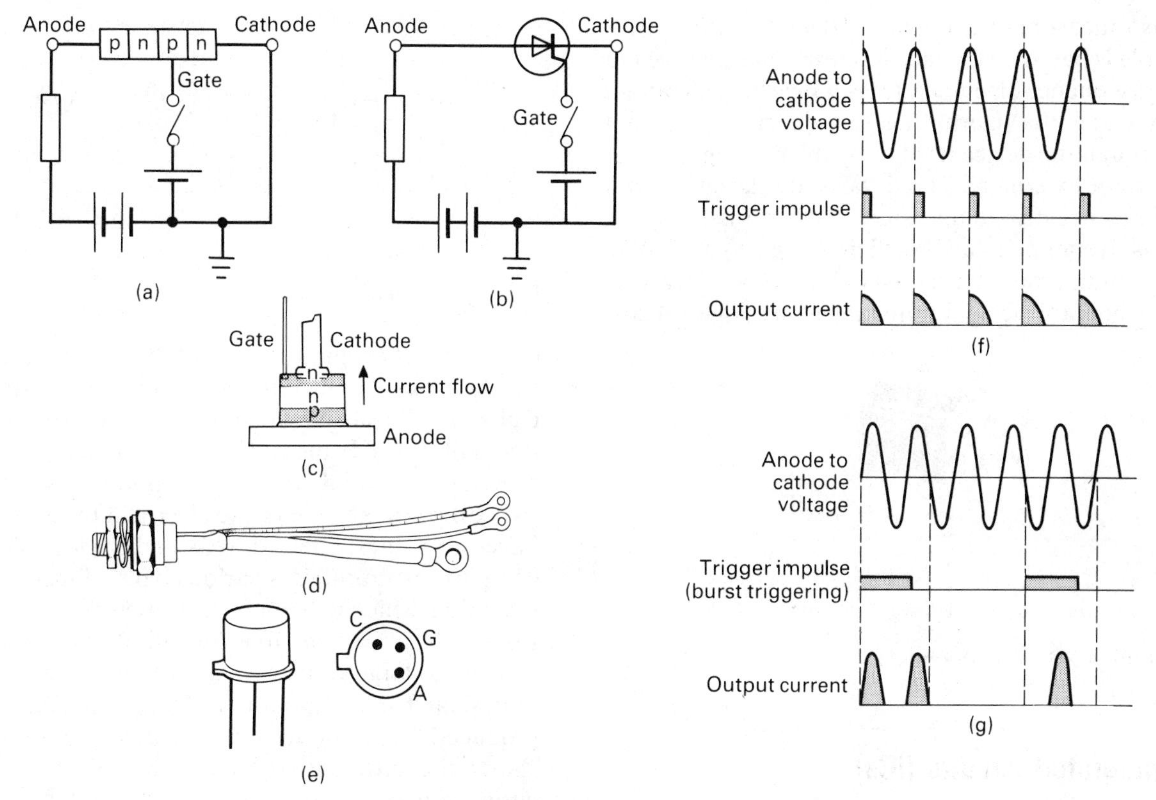

Figure 4.24 Structure, appearance, gate triggering and waveform of a thyristor

be controlled by altering gate voltage pulse times. It should be remembered that the thyristor can only conduct during positive half cycles, so gate triggering can be used to control power to loads. Figure 4.24 helps to explain this. Diagram (a) shows a small trigger voltage pulse (on the gate terminal) at time intervals. In graph (f) the left-hand corner edge of each pulse when traced up to the voltage signal only allows the positive quarter of the voltage waveform to conduct. This is seen in the similar portion of output current on the lower graph. In diagram (b) a more lengthy pulse is used and over different intervals, the output current to the load can be controlled (see graph (g)).

DIACS AND TRIACS

These are multijunction devices which are triggered into conduction in either their forward bias or reverse bias modes. A **diac** is a silicon bi-directional trigger diode, i.e. a device with two diodes connected in parallel with each other but in opposite directions. It is used in triac firing circuits, acting as an open switch until the applied voltage reaches about 35 V, when it will conduct. The triac is a gated bi-directional thyristor, i.e. a device with two thyristors connected in parallel but in opposite

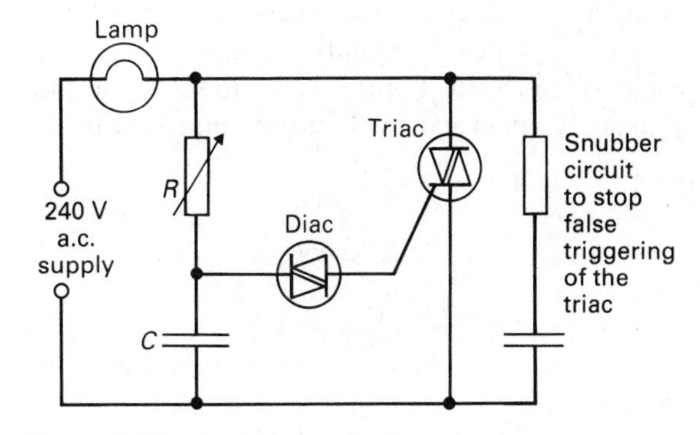

Figure 4.25 Basic lamp dimmer circuit

directions. This device can be triggered on both halves of each a.c. cycle, producing a full-wave d.c. output signal. By varying pulses to its gate, the power to the load can likewise be varied. Figure 4.25 shows how a diac and triac are connected in a simple lamp dimmer control circuit. They are widely used in control circuits.

THERMISTOR

This is a temperature-sensitive semiconductor resistor, which is designed for overtemperature protection in trip circuits and warning circuits, and also measurement circuits. Their operating principle is based on the fact that resistance change is a consequence of temperature change brought about by either the internal heating effect of a current through the device itself or by the heating effect of current externally. The devices are designed with either negative temperature coefficients (n.t.c.) or positive temperature coefficients (p.t.c.) and there are numerous types available today covering considerable temperature ranges (see Figure 4.26).

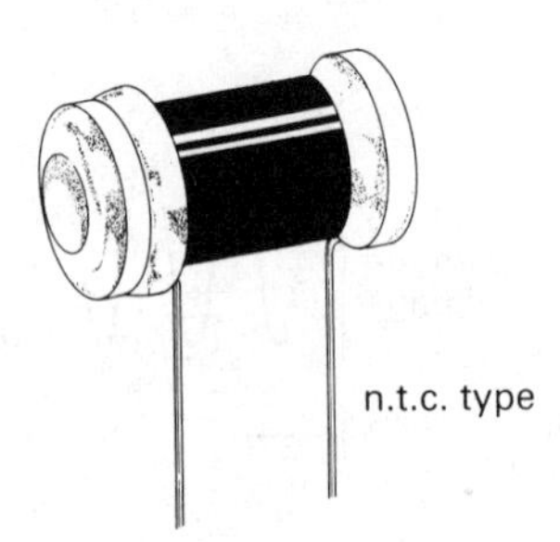

Figure 4.26 Rod thermistor

Integrated circuits (ICs)

These are miniaturised electronic packages comprising many of the components mentioned above and types of amplifier, connected together on a **silicon chip**. Figure 4.27 shows a common plastic IC package, with numerous dual-in-line

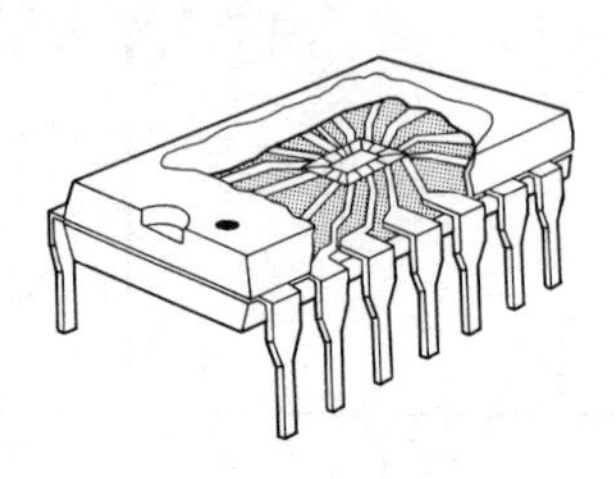

(a) Silicon chip

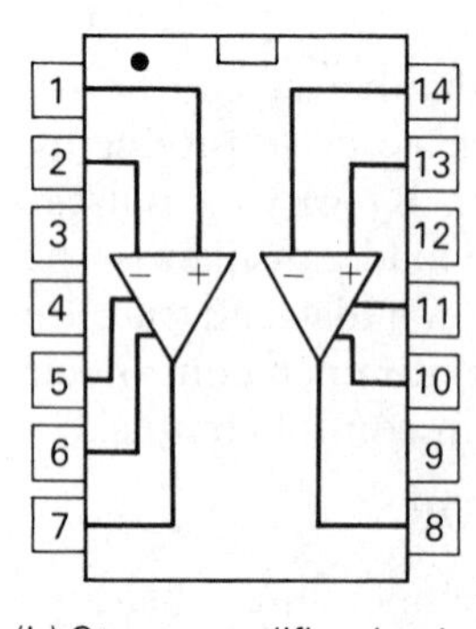

(b) Stereo amplifier circuit

Figure 4.27 IC package

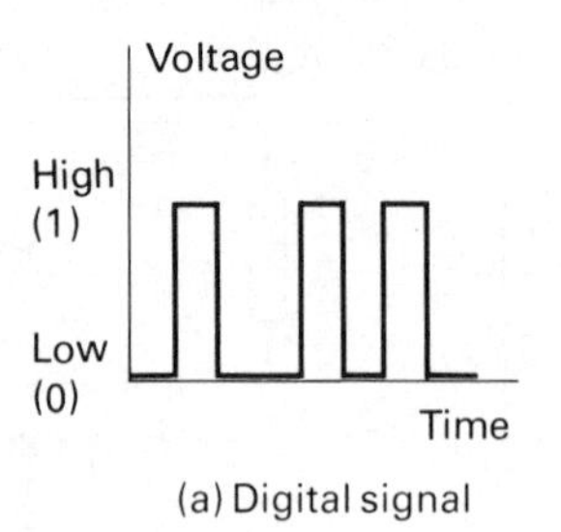

(a) Digital signal

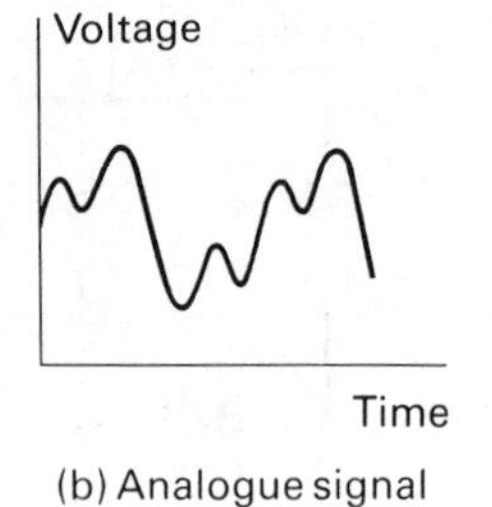

(b) Analogue signal

Figure 4.28 Digital pulses

(dil) connection pins attached to the 'chip'. There can be from 8 to 64 connection pins. ICs can be **digital** or **linear** in operation. Digital ICs are frequently used in pocket calculators, digital watches, etc. and these incorporate two-state switching circuits where the electrical pulses only have two values, referred to as high or low. Their complex operation uses various types of logic gate switching, which will not be discussed at this stage. Linear ICs are capable of amplifying small voltages to large voltages (say 1 μV to 1 V) and are represented by smooth analogue signals as produced by amplifier-type circuits. Figure 4.28 shows the difference between the two types of output signal.

Alternating current waveforms

These have already been mentioned in Chapter 3. Figure 4.29 shows the cyclic nature of a sinusoidal alternating current waveform. A waveform is a picture showing how current or voltage quantities change over a period of time. The quantity is usually represented on the vertical axis while frequency is represented on the horizontal axis. You will see that the waveform periodically reverses its direction in a regular pattern, from a positive value to a negative value. The fastest rate of change occurs when the waveform passes through zero. There are two peaks and these represent the **amplitude** or **peak values** of the waveform. One complete alternation is called a **cycle**. The time taken for one complete cycle is called the **periodic time** (T) and the number of cycles made in one second is called the **frequency** (f). Periodic time can be expressed as

$$T = \frac{1}{f} \qquad [4.1]$$

In SI Units, the unit of frequency is **hertz** (Hz), and 1 Hz = 1 cycle/sec.

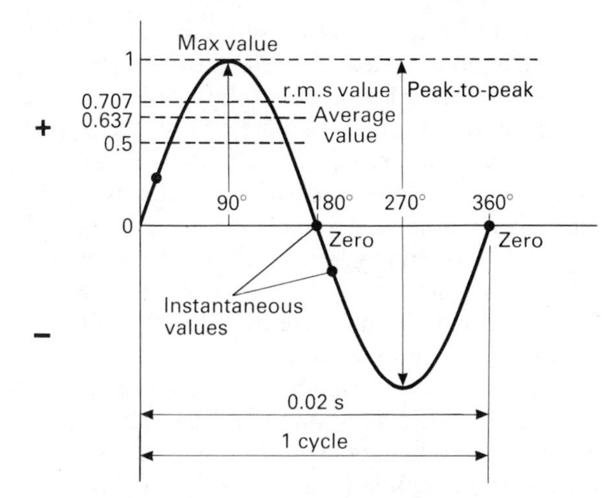

Figure 4.29 One cycle of a sinusoidal waveform at 50 Hz

The frequency of the alternating current we receive in our homes from the public supply is 50 Hz, but in radio, television and other electronic work, very high values are found. For example, the low frequency band for long wave radio ranges from 30 kHz to 300 kHz; medium frequency for medium wave is from 300 kHz to 3 MHz; high frequency for short wave is from 30 MHz to 300 MHz; and ultra high frequency for FM (frequency modulation) extends to 3 GHz. It should be noted that **audio frequencies** (a.f.) (i.e. sound waves created from loudspeakers) are produced over a range between 20 Hz and 20 kHz, whereas **radio frequencies** (r.f.) are from 20 kHz upwards.

There are other terms associated with the sine wave. For example, the peak value is often called the **maximum value**. Since both positive and negative halves of the waveform are equal, the **average value** is zero. The average value of half a sine wave is found to be **0.637** times the maximum value. All a.c. supplies use the measurement called **root-mean-square value** (r.m.s. value) and this is **0.707** times the maximum value. This value is the equivalent value of d.c. current that would produce the same heating effect in a pure resistive component. Any point on the sine wave at any instant is called an **instantaneous value** (v). This value can be found easily if you know the maximum value of the sine wave (V_{max}) and the number of degrees it is from zero (ϕ). The following formula is used:

$$v = V_{max} \sin \phi \qquad [4.2]$$

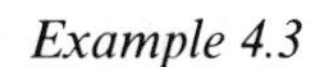

Example 4.3

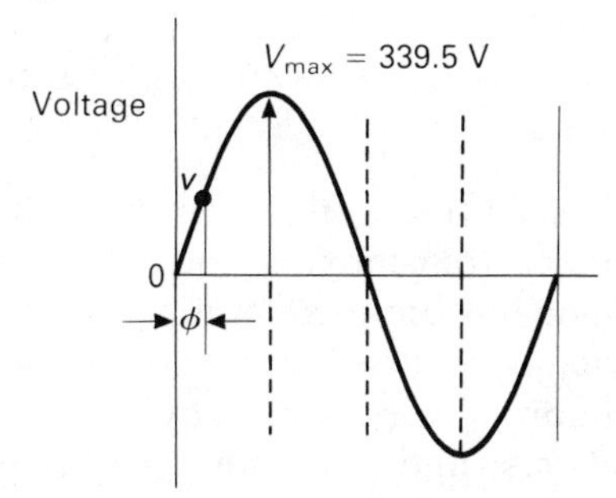

Figure 4.30 Waveform of voltage quantity

Figure 4.30 shows one cycle of a 50 Hz sinusoidal waveform voltage having a maximum value of 339.5 V. If the phase angle (ϕ) shown is 10°, determine the following:

a) the instantaneous value;
b) the average value;
c) the r.m.s. value;
d) the time taken to reach 10° and 180°.

Solution

a) Using formula [4.2]

$$v = V_{max} \sin 10°$$
$$= 339.5 \times 0.1736 = 58.94 \text{ V}$$

b) Average value

$$V_{ave} = 0.637 \times V_{max}$$
$$= 0.637 \times 339.5 = 216.26 \text{ V}$$

c) r.m.s. value

$$V_{r.m.s.} = 0.707 \times V_{max}$$
$$= 0.707 \times 339.5 = 240 \text{ V}$$

d) Using formula [4.1]

$$T = \frac{1}{f}$$

for 50 Hz

$$T = \frac{1}{50} = 0.02 \text{ s}$$

Time taken to reach 10°

$$t_1 = 0.02 \times \frac{10}{360} = 0.000\,555 \text{ s}$$

Time taken to reach 180°

$$t_2 = 0.02 \times \frac{180}{360} = 0.01 \text{ s}$$

In electronic work, when components such as rectifiers, transformers and other mixed circuit components are used, **distortion** of the sine wave inevitably occurs. Figure 4.31 shows some of the common shapes that may be obtained from **oscillators** and **waveform generators** (electrical equipment for testing radio sets, etc.) particularly for measurement reasons. Distortion is the result of interference from **harmonic currents**, which are sine waves of different frequencies, amplitudes and even phase difference. Some of these are shown in Figure 4.32(a). When harmonics mix with the **fundamental** sine wave, a **complex wave** is created. Figure 4.32(b) shows how 2nd and 3rd harmonic currents of smaller amplitude cause distortion of the fundamental sine wave and create their own individual complex waves.

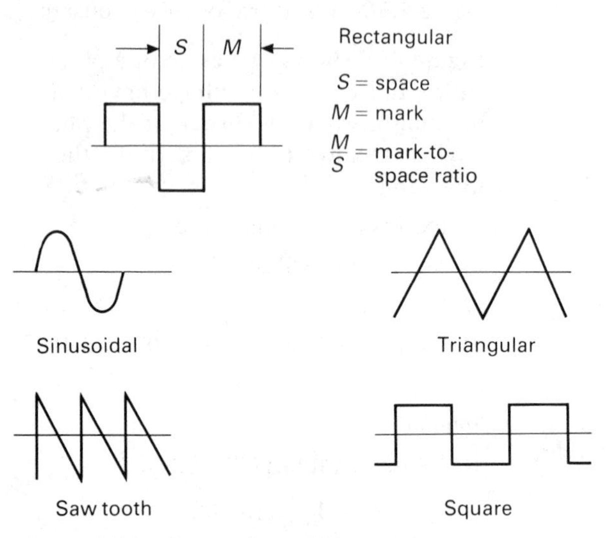

Figure 4.31 Types of waveform

The triangular waveform in Figure 4.31 is created from the fundamental sine wave and a mixture of other **even** harmonic waves. This waveform increases linearly in both positive and negative directions. The saw tooth waveform is a particular type of triangular waveform. Both types of triangular waveform are used where time measurement is required, for example, in television monitors, oscilloscopes and for timing radar pulses.

The square wave is a particular type of rectangular wave, where the quantity being measured changes instantaneously from both peaks. It is made up of the fundamental wave and a mixture of **odd** harmonics. In these types of waveform the **mark-to-space ratio** provides information on how the speeds of two switching processes can be compared on say an oscillator or oscilloscope.

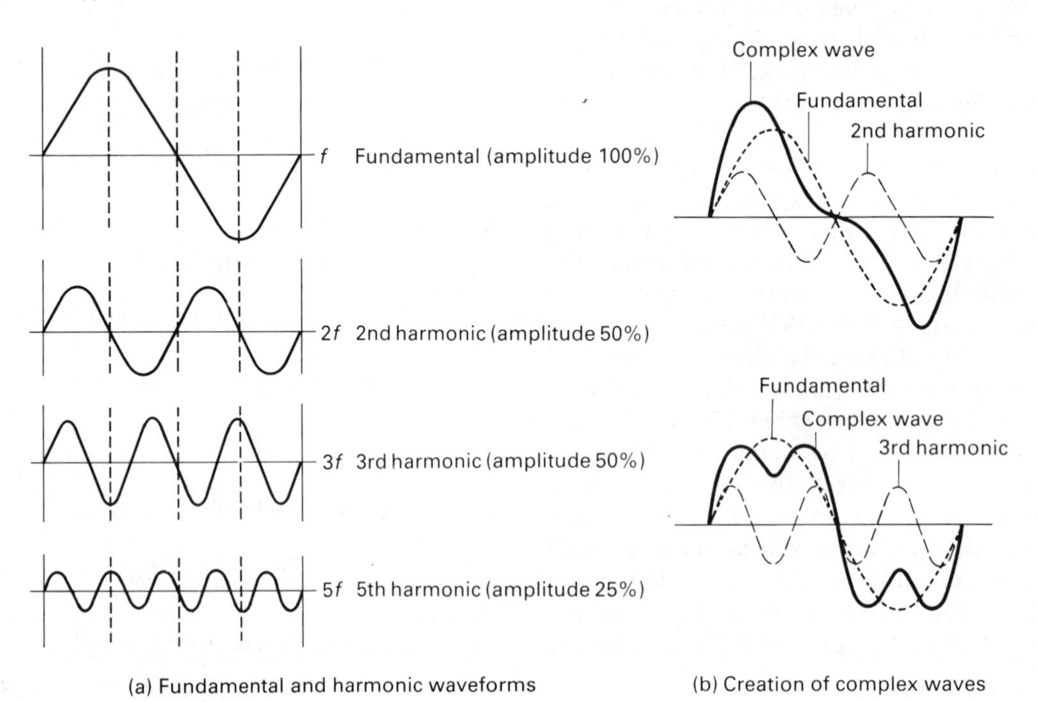

Figure 4.32 Waveform distortion

When this ratio equals 1 the rectangular waveform produced is a square wave. **Pulse trains** are another type of waveform produced for measurement in oscillators, digital computers and other information systems (see Figure 4.28).

EXERCISE 4

1. a) Figure 4.33(a) shows a carbon composition resistor of 15 MΩ with 10% tolerance. What are its band colours and what are its possible higher and lower values?
   b) Figure 4.33(b) shows a polyester capacitor. From its band colour markings, determine its capacitance value in picofarads and microfarads, and also its tolerance and voltage values.

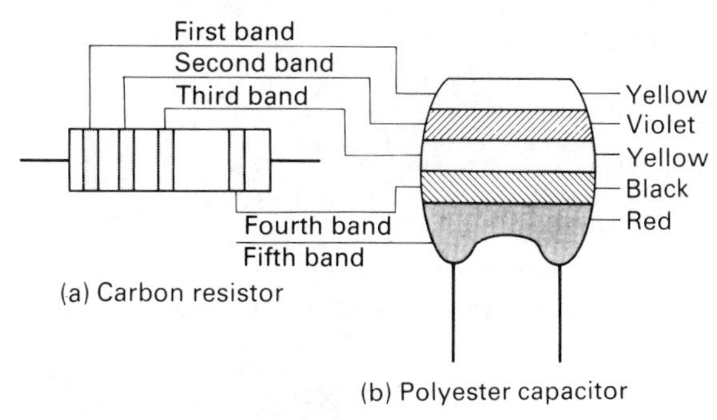

Figure 4.33 Colour codes

2. Briefly state the purpose of the following components:
   a) variable resistor;
   b) variable capacitor;
   c) zener diode;
   d) transistor.

3. a) What is meant by the term **stability** of an electronic component?
   b) Name one type of capacitor that is polarised and one that is non-polarised.
   c) Why do polarised capacitors have to be connected correctly?
   d) Briefly explain the function of the **gate** connection on a thyristor.

4. a) Which of the diagrams in Figure 4.34 represents an iron core transformer? Give an example where it could be used.
   b) What is the purpose of a transformer, and why are iron-cored transformers laminated?

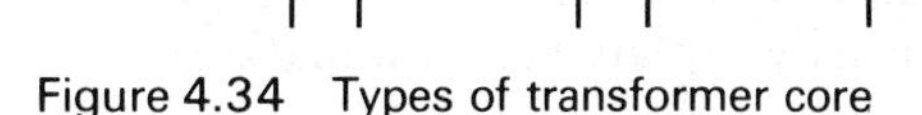

Figure 4.34 Types of transformer core

5. a) Make a diagram of a p–n junction diode and explain how forward and reverse bias occurs.
   b) Figure 4.35 shows two 2 V/40 mW lamps connected in series across a 3 V supply. When switch S is closed, what is the condition of each lamp: bright, dim or off?
   c) If the diode connection were reversed, what would happen?

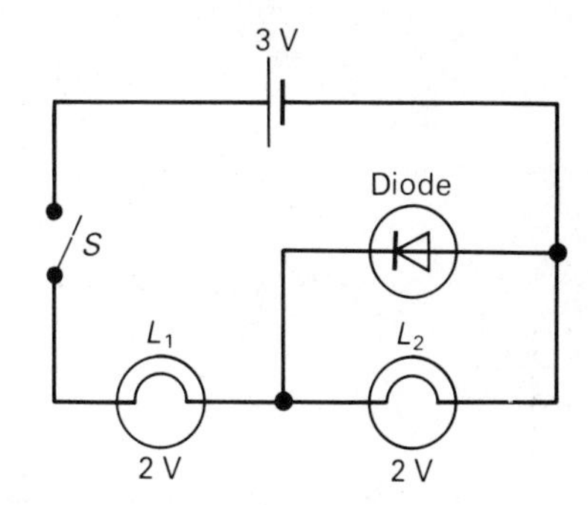

Figure 4.35 Lamp control circuit

6. a) Show, by neatly labelled diagrams, two methods of connecting a transistor in a circuit.
   b) State the modes of connection in a) above.
   c) Name two applications involving the connection of transistors.

7. a) Show, in a neatly labelled diagram, the three connections on a thyristor.
   b) Explain the function of the device on a.c. and d.c. supplies.
   c) State two applications involving use of a thyristor.

8. Draw a neatly labelled circuit diagram indicating how full-wave rectification is achieved from two semiconductor diodes and a centre-tapped transformer. Make a sketch of the output signal.

9. a) A sinusoidal waveform voltage over half a cycle has an r.m.s. value of 100 V. What are its peak and average values?
   b) Sketch two cycles of a square waveform and briefly explain the term mark-to-space ratio.

10. Figure 4.36 shows a current sine wave with a maximum value of 10 A.
    a) What is its average value and peak-to-peak current?
    b) If the duration of one cycle is 0.0166 s, what is its frequency?
    c) What are the points marked A, B and C?
    d) If ϕ is 120° at C, what is the value of current at this point?

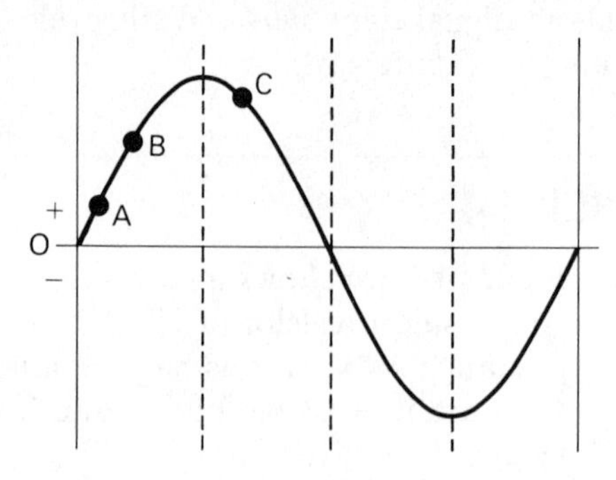

Figure 4.36 Current waveform

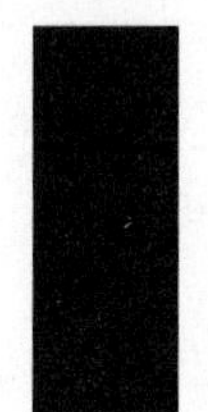

Appendix 1

Multiple choice questions

Objective

After reading this book you should be able to complete the following multiple choice questions.

TERMINOLOGY

1. The unit of mass is called a

A newton C pascal
B kilogram D tonne

2. The freezing point of water (0 °C) is equivalent to

A 100 K C 212 K
B 173 K D 273 K

3. The mass per unit volume of a substance is called its

A matter C density
B pressure D weight

4. Convert 600 mA to amperes

A 0.006 A C 0.6 A
B 0.06 A D 6 A

5. The symbol for the metric prefix 'micro' is

A μ C M
B m D k

6. The number 10^0 is equal to

A 0.1 C 10
B 1.0 D 100

7. One joule of energy is equivalent to

A 1 newton metre C 1 millivolt
B 1 ohm metre D 1 kilohertz

8. The quantity of electricity is found from the formula

A $Q = I \times t$ C $Q = P \times t$
B $Q = V \times I$ D $Q = I \times R$

9. The Greek letter ρ is the symbol for

A resistance C resistivity
B reactance D reluctance

10. The area of a circle is found by the formula

A πr C $\pi d/4$
B πr^2 D $2\pi r$

11.

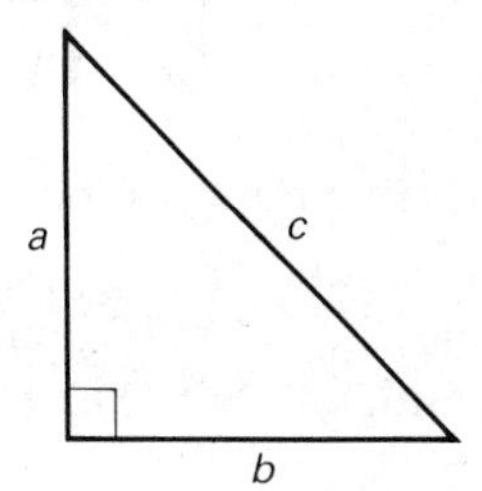

In the right-angled triangle shown, the square on the hypotenuse side (c) is equal to

A $a^2 - b^2$ C $a^2 \times b^2$
B $a^2 + b^2$ D $a^2 \div b^2$

12.

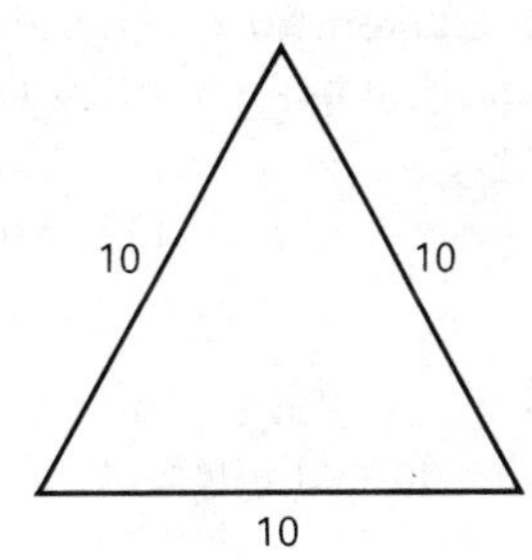

What is the approximate area in square units of the triangle shown?

A 15 C 43
B 30 D 50

13. A double-wound transformer has a ratio 10 : 6. If the primary winding is supplied at 250 V and the transformer has 60 secondary turns, what is the secondary voltage and the number of primary turns?

A 415 V and 150 T C 30 V and 415 T
B 100 V and 30 T D 150 V and 100 T

14. What percentage of 415 V is 10.375 V?

A 2.5% C 6.0%
B 5.0% D 7.5%

15. Which of the following has the highest melting point?

A silver C aluminium
B gold D tungsten

16.

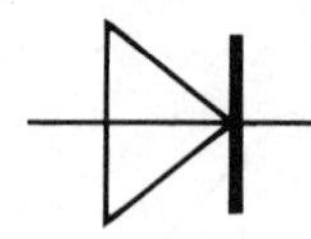

The BS3939 circuit symbol shown is a

A semiconductor diode C supply metering point
B amplifier D tap-change transformer

17. The third band colour of a 25 kΩ radio resistor is

A red C yellow
B orange D green

18. The time taken for one cycle (1 Hz) to flow in a supply generated at 50 Hz is

A 0.02 s C 50 s
B 1.00 s D 100 s

19. The property of an inductor or capacitor to resist the flow of alternating current is called

A resistance C reactance
B impedance D reluctance

20.

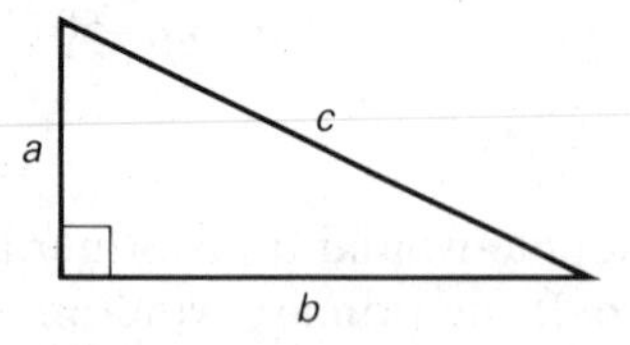

In the diagram, if $a = 0.03$ and $b = 0.04$, what is c?

A 0.01 C 0.05
B 0.02 D 0.06

MECHANICAL SCIENCE

21. The force needed to stop and start things is called

A friction force C acceleration force
B centripetal force D inertial force

22. The force of gravity on a 6 kg mass is approximately

A 1 N C 60 N
B 6 N D 100 N

23. A force of 100 N is required to accelerate a mass of 5 kg at

A 5 m/s^2 C 100 m/s^2
B 20 m/s^2 D 500 m/s^2

24. All the following are vector quantities, **except**

A speed C velocity
B force D acceleration

25.

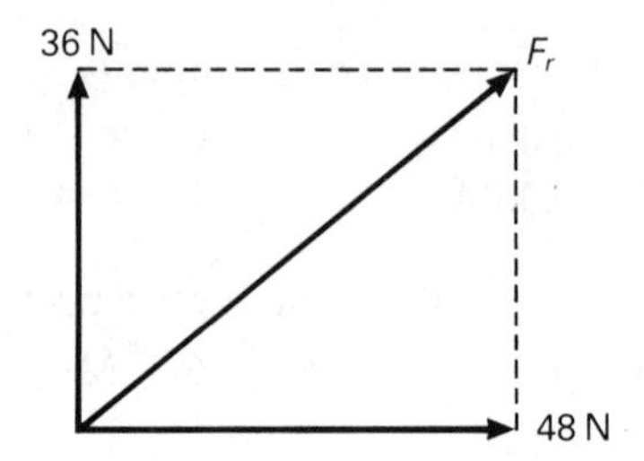

In the diagram, what is the resultant force?

A 42 N C 84 N
B 60 N D 100 N

26.

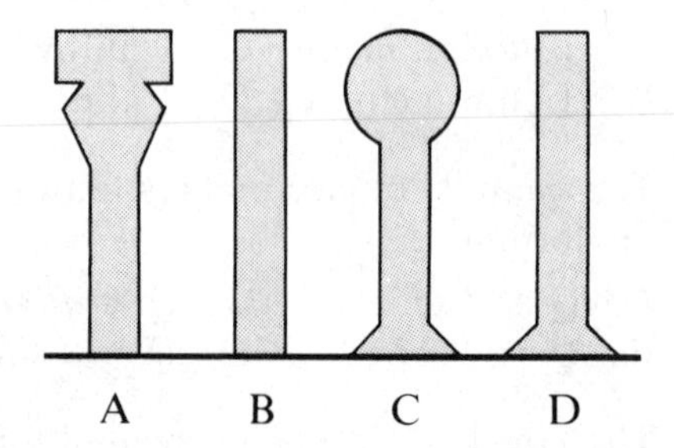

Which of the free-standing objects in the diagram has the more stable equilibrium?

27.

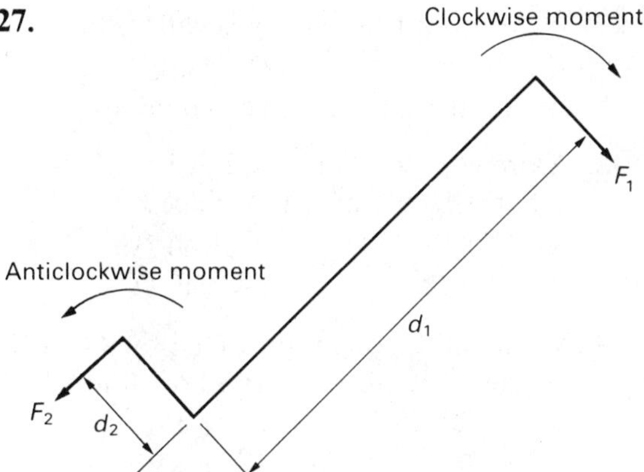

The diagram illustrates the principle of moments. What approximate effort F_1 has to be applied to lift a load of 60 kg if $d_2 = 100$ mm and $d_1 = 0.4$ m?

A 150 N C 400 N
B 200 N D 600 N

28. What is the density of an object with a mass of 100 g and a volume of 20 cm^3?

A 5 g/cm^3 C 100 kg/cm^3
B 2 kg/cm^3 D 120 kg/cm^3

29. A box weighs 200 N and its base has measurements of 0.5 m × 1.5 m. What approximate pressure does it exert on the ground?

A 150 N/m^2 C 267 N/m^2
B 203 N/m^2 D 300 N/m^2

30. What is 86 °F converted to °C?

A 25 °C C 42 °C
B 30 °C D 50 °C

31. An aluminium pipe 1 m long has a linear coefficient of expansion of 0.000 026 per °C. What is its expansion when heated to 80 °C?

A 0.208 mm C 0.208 dm
B 0.208 cm D 0.208 hm

32. The amount of heat needed to raise the temperature of 1 kg of substance by 1 °C is called

A specific heat density C specific heat capacity
B specific latent heat D specific gravity heat

33. What is the power of a crane lifting a load weighing 2000 N through a height of 30 m in 15 s?

A 220 W C 10 kW
B 4 kW D 100 kW

34. The part of a circle that force has to be multiplied by to produce torque is called

A circumference C circle constant (pi)
B diameter D radius

35. The ratio $\frac{\text{Load}}{\text{Effort}}$ is called

A efficiency C mechanical advantage
B velocity ratio D friction force

36. What is the approximate velocity ratio of the screwjack shown below, if the screw has 500 threads per metre and the handle length is 0.191 m?

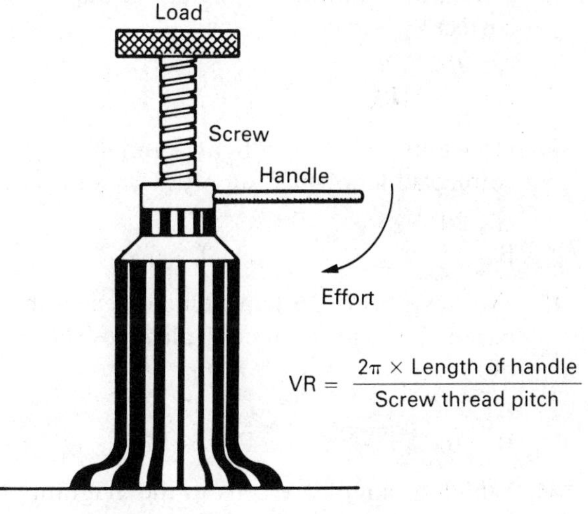

A 100 C 450
B 300 D 600

37. The type of force created by two surfaces moving in opposite directions is called

A tensile force C cohesive force
B compressive force D friction force

38. What is 500 degrees Celsius converted into degrees Fahrenheit?

A 300 °F C 900 °F
B 332 °F D 932 °F

39. To make a thermometer 'quick acting', it should have a bulb glass which is

A thin C long
B thick D short

40. Which of these substances has the lowest specific heat capacity?

A paraffin C water
B mercury D copper

ELECTRICAL SCIENCE

41. Which particle of an atom has a negative charge?

A proton C electron
B neutron D ion

42. At normal temperatures, all of the following are good conductors of electricity, **except**

A lead C glass
B iron D tin

43. Which formula for finding current is correct?

A $I = VR$ C $I = V^2R$
B $I = V/R$ D $I = V^2/R$

44. How many 12 V fairy lights can be connected to a 240 V supply?

A 20 C 12
B 15 D 8

45. Two resistors of 16 Ω, when connected in parallel, will have an equivalent resistance of

A 8 Ω C 32 Ω
B 16 Ω D 64 Ω

46. Which group of 2 V cells in the diagram provides the highest voltage output?

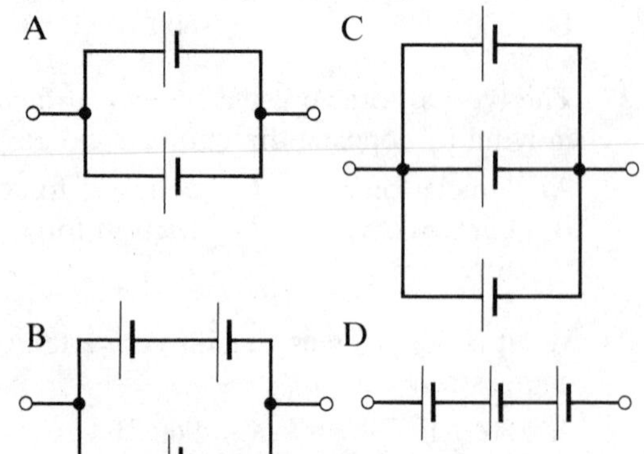

47. The unit of heat energy is called the

A kilowatt C joule
B ohm metre D coulomb

48. A projector lamp is rated at 1.5 kW/240 V. What is its working resistance?

A 6.25 Ω C 160 Ω
B 38.4 Ω D 240 Ω

49. What is the power consumed by a 60 kΩ resistor if it takes a current of 10 mA?

A 6 μW C 6 kW
B 6 W D 6 MW

50. A consumer uses 18 MWh of electricity. What is the unit charge at 6 p/unit?

A £108 C £10 800
B £1080 D £108 000

51. A 3 kW/240 V load takes a current of 12.5 A at unity power factor. What current is taken at 0.5 power factor?

A 6 A C 25 A
B 13 A D 50 A

52. Which type of component has unity power factor?

A motor windings C capacitor
B fire-bar element D discharge lamp ballast

53. Which of the following is the ratio for efficiency?

A $\frac{\text{Work output}}{\text{Work input}}$ C $\frac{\text{Heat input}}{\text{Heat produced}}$

B $\frac{\text{Energy in}}{\text{Energy out}}$ D $\frac{\text{Work done}}{\text{Power output}}$

54. The diagrams show the magnetic field pattern for two conductors in close proximity with each other. Which is correct?

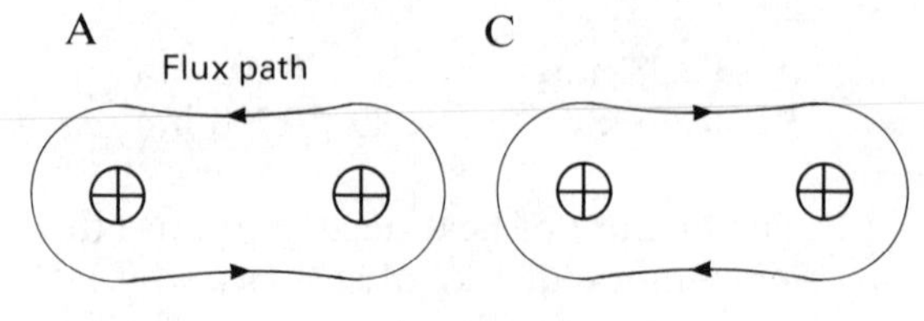

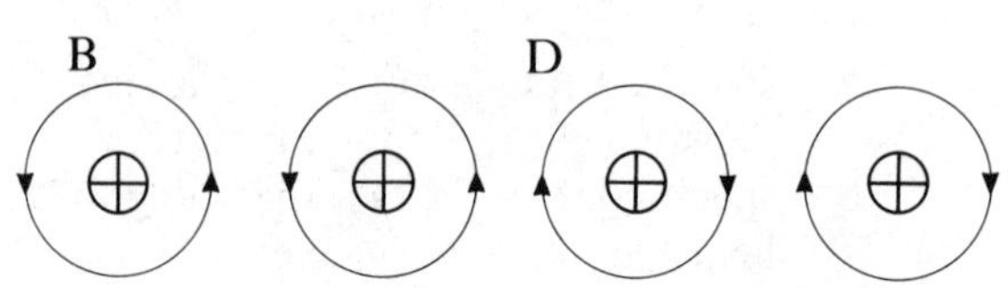

55. All the following may alter the strength of induced e.m.f. **except**

A velocity of the moving conductor
B length of conductor in magnetic field
C strength of magnetic flux density between poles
D cross-sectional area of conductor in magnetic field

56. Which type of supply produces a sinusoidal waveform?

A secondary battery
B d.c. generator
C a.c. alternator
D solar cell

57. Fleming's right hand rule is a method of finding the direction of

A force on a conductor
B induced electromotive force
C force of gravity
D magnetomotive force

58. Which of these devices works on the principle of electromagnetic induction?

A transformer
B rectifier
C capacitor
D transistor

59.

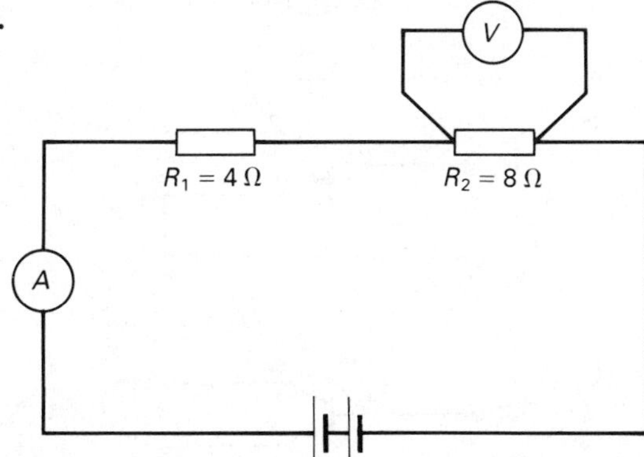

In the circuit shown, the reading on the ammeter is 5 A. What is the reading on the voltmeter?

A 10 V
B 20 V
C 40 V
D 60 V

60.

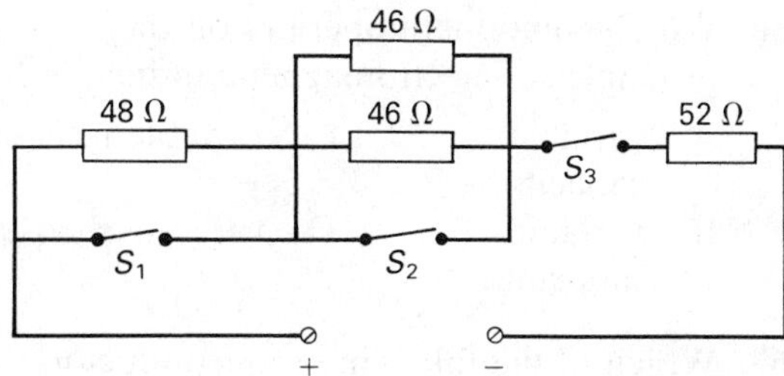

To obtain a resistance of 75 Ω in the circuit shown the position of all switches should be:

A S_1 off, S_2 on, S_3 on
B S_1 on, S_2 on, S_3 off
C S_1 off, S_2 on, S_3 off
D S_1 on, S_2 off, S_3 on

BASIC ELECTRONICS

61. The third colour band on a carbon radio resistor is the

A unit number
B tens number
C tolerance value
D number of 0s

62. The fifth colour band on a polyester capacitor is the

A working voltage
B tolerance value
C number of 0s
D last number

63.

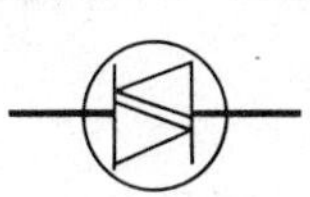

The BS3939 symbol shown is a

A thyristor
B transistor
C diac
D triac

64. The printed code for a radio resistor of value 16.5 kΩ is

A 16K5
B 165K
C 16R5
D 165R

65.

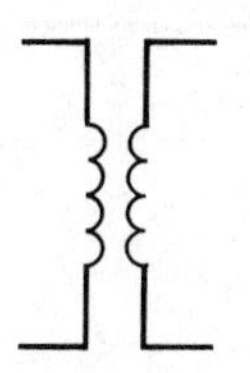

What is the type of transformer shown?

A iron-cored
B air-cored
C ferrite dust cored
D isolated cored

66. Which component operates on the principles of electromagnetic induction?

A variable capacitor
B variable inductor
C variable resistor
D variable rheostat

67. Which of the following semiconductor devices has a gate connection?

A diode
B transistor
C diac
D thyristor

68. All these colours are produced by a light emitting diode, **except**

A red
B yellow
C green
D white

69. Which of the following is a semiconductor material?

A germanium
B copper
C brass
D gold

70. Which of the following capacitors **must** be connected correctly?

A polyester
B polycarbonate
C electrolytic
D mica

71. The output waveform produced in the circuit shown is

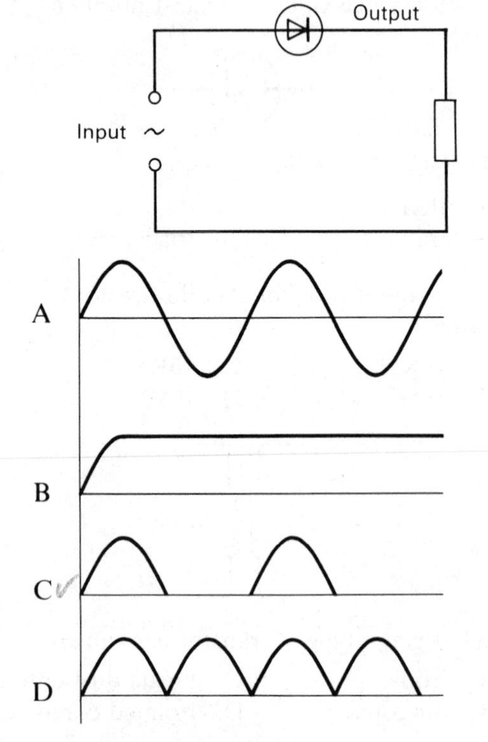

72.

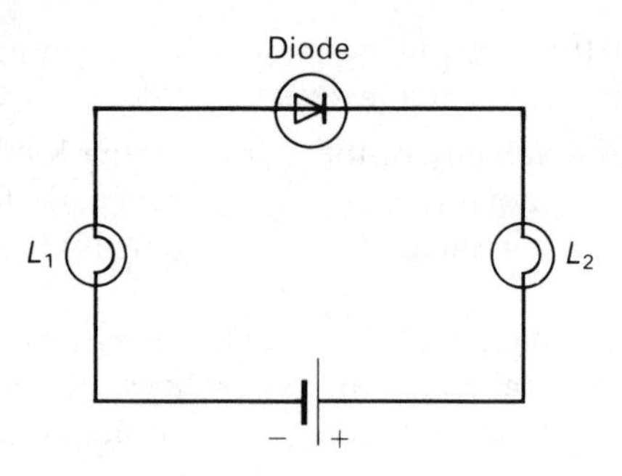

The diode shown is connected to create

A reverse bias
B forward conduction
C negative polarity
D dim light

73. A heat sink is a device which

A acts as a support for a soldering iron when heated
B allows heat to be transmitted away from a semiconductor device
C contains waste solder after it has been used on a joint
D acts as trip mechanism inside a thermistor at hot temperatures

74. Which of the following connections is correct for a common emitter npn transistor?

A
In Out — b; e — In; c — Out

B
In — b; e — Out; c — In Out

C

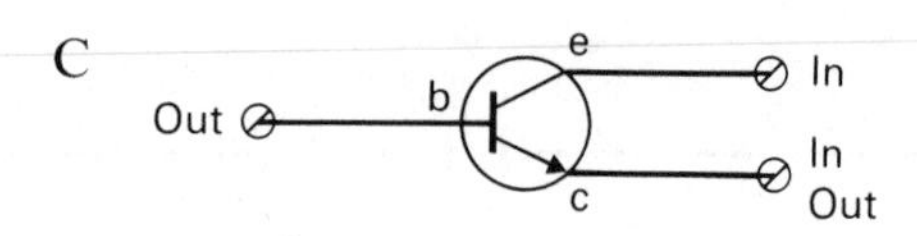

D

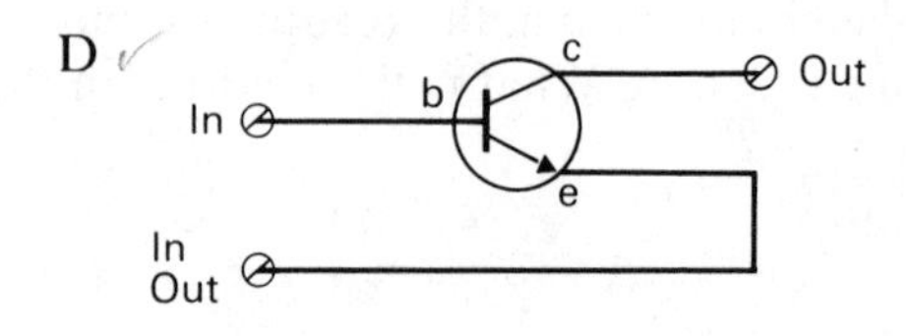

75. Which of the following circuits change a.c. to d.c.?

A rectifying circuit C meter protection circuit
B voltage reference circuit D logic gate circuit

76. The periodic time for a supply frequency of 100 Hz is

A 0.1 s C 0.001 s
B 0.01 s D 0.0001 s

77. The maximum value reached by a sinusoidal a.c. supply voltage of 50 V r.m.s. is

A 35.35 V C 78.49 V
B 70.72 V D 86.60 V

78. A square wave is produced from

A a number of odd harmonics waves plus fundamental
B a number of even harmonics waves plus fundamental
C a clean a.c. supply without harmonics waves
D a mixture of odd and even harmonics waves

79. When a single thyristor on a.c. conducts it produces

A full-wave rectification C distorted rectification
B half-wave rectification D saw tooth rectification

80. Which of these BS3939 symbols is a light sensitive diode?

A
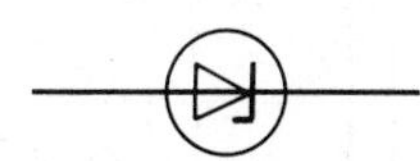

B
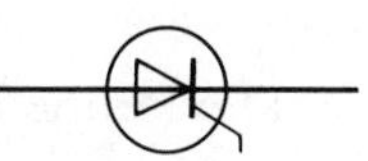

C
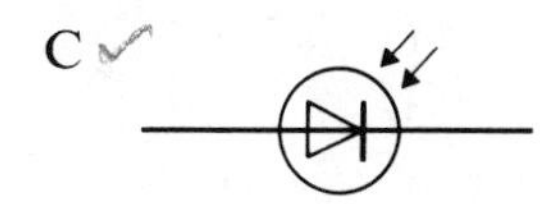

D
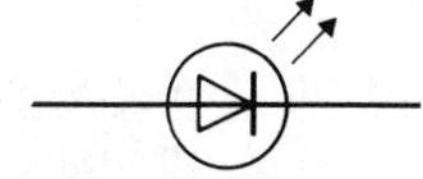

GENERAL

81.

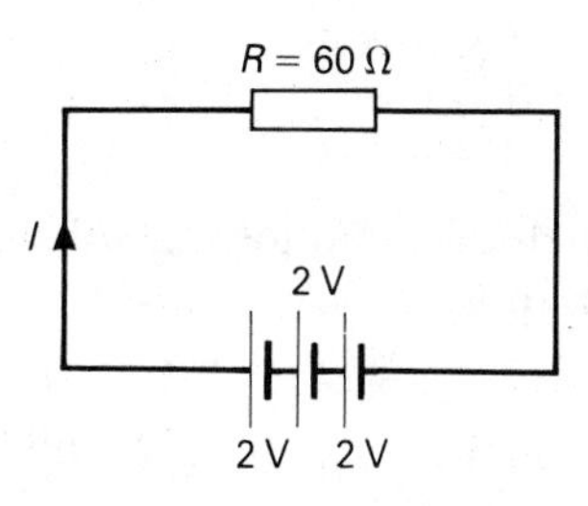

The current taken by the circuit shown is

A 0.1 mA C 10 mA
B 1.0 mA D 100 mA

82. The unit of weight is the

A kilogram C newton
B pascal D coulomb

83. The henry is the unit of

A capacitance C reactance
B impedance D inductance

84. A wattmeter measures

A energy C power factor
B power D voltamperes

85. The negative conductor colour identifying a d.c. two-wire supply is

A black C blue
B white D purple

86.

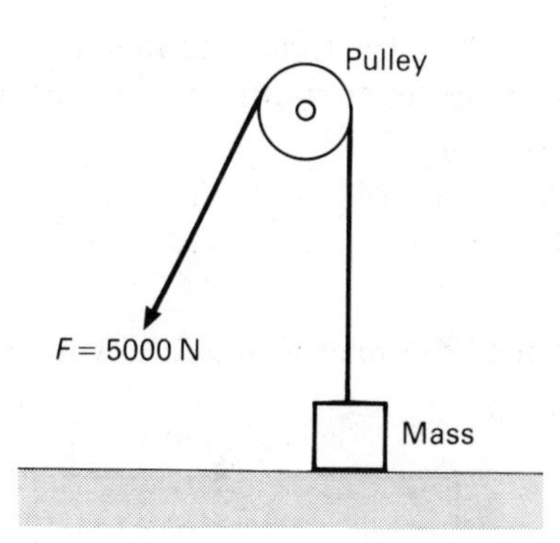

Taking g as 10 m/s^2, the mass lifted by the pulley shown is

A 500 mg C 500 kg
B 500 g D 500 Mg

87. What force has to be applied to a mass of 29.5 kg to accelerate it at 0.2 m/s^2?

A 1.5 N C 9.81 N
B 5.9 N D 14.75 N

88. Which of the following substances will sink in water?

A ice C air
B petrol D mercury

89. The ratio $\frac{\text{Force}}{\text{Area}}$ is called

A pressure C efficiency
B density D moment

90. Which of the anticlockwise moments shown will create equilibrium?

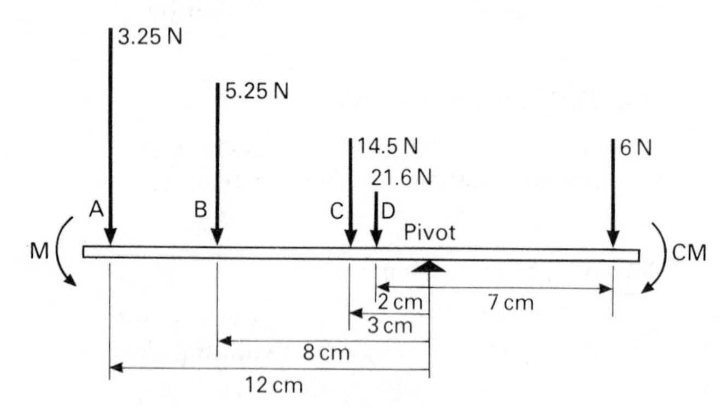

91. The negatively-charged particles in orbit around the nucleus of an atom are called

A electrons C protons
B neutrons D ions

92. Which of the following materials has a negative temperature coefficient of resistance?

A carbon C copper
B silver D aluminium

Questions 93–5 refer to the following diagram.

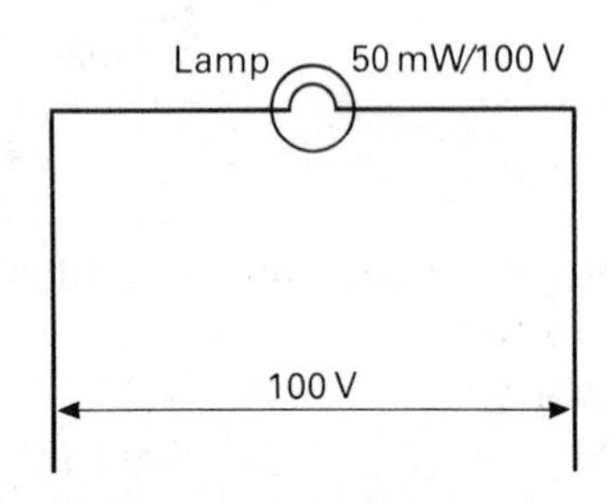

93. What is the working resistance of the lamp shown?

A 200 μΩ C 200 kΩ
B 200 mΩ D 200 MΩ

94. If another 50 mW/100 V lamp were placed in parallel with the lamp, the equivalent circuit resistance would be

A 100 μΩ C 100 kΩ
B 100 mΩ D 100 MΩ

95. If another 50 mW/100 V lamp were placed in series with the lamp, each lamp's power would be reduced by

A one twelfth C one sixth
B one eighth D one quarter

96. What energy is taken in a 240 V circuit for a charge of 50 kC?

A 12 μJ C 12 kJ
B 12 mJ D 12 MJ

97.

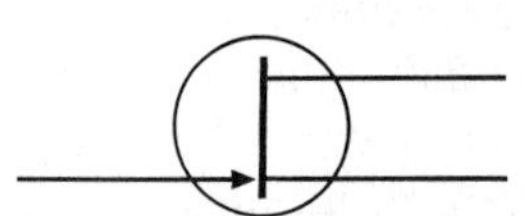

What is the name of the semiconductor device shown?

A field effect transistor C thyristor
B bipolar transistor D zener diode

98.

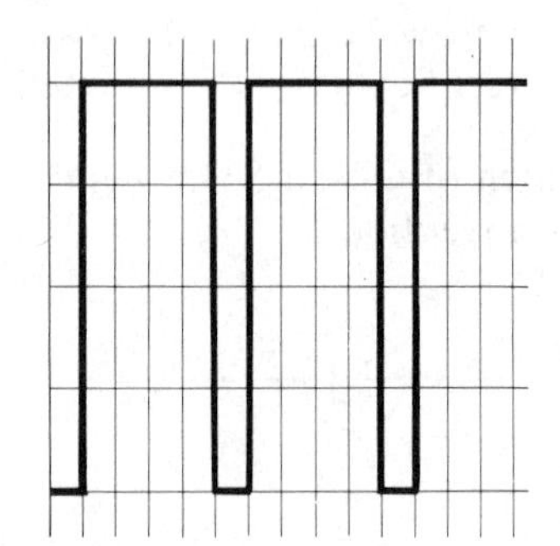

The mark-to-space ratio for the rectangular wave shown is

A 1 C 3
B 2 D 4

99. What is the value of a radio capacitor marked 1m?

A one microfarad C one megafarad
B one millifarad D one macrofarad

100. In the diagram below the horizontal line represents the root mean square value.

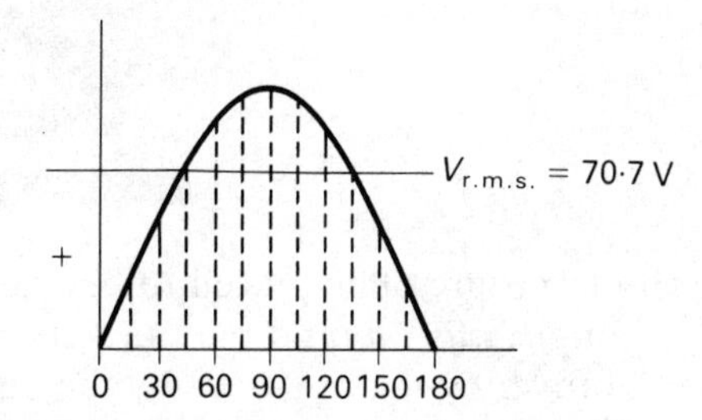

What is the maximum value if it is sinusoidal?

A 90 V
B 100 V
C 240 V
D 415 V

Appendix 2

Short-answer questions

Objective

After reading this book you should be able to complete the following sentences correctly.

TERMINOLOGY

1. The unit of length is the m .... and the unit of mass is the k .... They are both SI Units but other units, like the w ... (the unit of p .... and defined as the dissipation of energy) are d ...... units.

2. In the metric system, p .... are used to simplify lengthy numbers associated with units. For example, 1000 can be represented by the single letter k meaning k ... and 1/1 000 000 can be represented by the Greek letter μ meaning m .... 0.0001 farads can be written as ..... and 20 000 000 ohms can be written as ....

3. Prefix letters can be expressed as powers of ten, for example 1 000 000 can be written as ... and 1/100 written as ...

4. The method of finding an unknown term in a formula is called t........... and there are certain rules to obey. A term with a plus sign on one side of an equal sign, will become a term with a m.... sign when it is moved across to the other side of the equal sign. The line which divides a numerator from its d.......... is called the q....... line. Single terms or grouped terms (terms attached to each other with a plus or minus sign) can move across the equal sign from bottom to top or vice versa and square root signs can vanish, provided all terms on the opposite side are s...... For example, $\sqrt{81} = 9$ and $81 =$ ..

5. Group terms are often placed in brackets, for example, $a(a+b)$ and the brackets can easily vanish by m.......... by the outside term. The above becomes $a^2 + ab$.

Remove the brackets from

(i) $a(2a - 4b)$ (ii) $(a+b)(a-b)$

Answers here:

(i) (ii)

6. In mensuration (the method of finding lengths, areas and volumes, etc.) π (pi) is a c....... for a circle and is the ratio of c............/d....... Its value 3.142 will be the same for any circle. The area (A) of a circle is found by the formula $A =$ ...

The formula for the volume (V) of a cylinder is $V =$ ..

7. In a right-angled triangle, the longest side is called the h......... and its square is equal to the s.. of the s...... on the o....... s....

8. There are .. degrees in a right angle, ... degrees in a triangle and ... in a circle. In a square there are four r.... ...... and the d....... line is longer than the sides by a factor of .....

9. A random point chosen on an a.c. waveform is called an i........... point and the value producing the same heating effect equivalent to d.c. is called the r... m... s..... value, which has a numerical value of ..... for a sinusoidal waveform.

10. Most conductor resistances are p........... to length, but their cross-sectional areas are i........ p........... If a conductor's length is doubled and its c.s.a. halved, its resistance will be f... times greater. If its length is halved and c.s.a. doubled its resistance will be f... times smaller.

MECHANICAL SCIENCE

1. The amount of matter contained in a body is called its m... A g... is 1000th of a k....... Force is measured in n...... and can take many different forms. For example, the force needed to stop and start things is called i....... force and that which places a material under tension is called t...... force. The force that acts downwards towards the Earth is called g........... force.

2. W..... is a force measured in newtons and any mass influenced by it has to be multiplied by a factor of .... This is the a.......... of f... f... If a bag of onions had a mass of 5 kg, it would weigh approximately .. N.

3. A vector quantity can be represented by its m........ and direction and if two forces act in different directions the r........ can be found by the method known as p........... of forces.

4. When two forces just balance each other they are said to be in e.......... For an object to be very stable, it should have a low c..... of g...... Balancing a ladder in the middle while carrying it on your shoulder is a good example of the p........ of m...... Here the c........ m..... should equal the a........... m.....

5. Density is m... per unit v..... and it varies with t.......... The term symbol is .. At 4 °C the relative density of water is .... kg/m^3. This is a useful way of comparing substances to see if they will f.... or s... The formula for finding r....... density is the ratio of density of substance to density of water.
   The mass of air required in a room of volume 90 m^3, given the density of air is 1.3 kg/m^3, is .....

6. Heat is a form of e..... but temperature is a state of h...... or c....... of a substance and is measured in d...... The liquid in a thermometer could be m...... or a ......
   The boiling point of water on the C...... scale is ....., and on the F......... scale it is ..... The K..... scale is the scale used for t........... t.......... and at absolute freezing point, 0 K is equivalent to .... °C.

7. Pressure is force per unit area. A force a..... over a small area gives a l.... pressure. Atmospheric pressure is approximately .... For a fixed mass of g.. at constant volume, pressure is directly proportional to the a....... t..........

8. Torque is t...... e..... or force times d....... Its unit is the n..... m....
   The mechanical power of a motor is expressed as P = ... watts.

9. Load divided by effort is called m......... a........ and this provides information on how e... or h... it is to move an object. Another useful term is called the v....... r.... which compares the d........ moved by effort and load.

10. If a pulley system has a VR of 3 and a load of 200 N requires an effort of 100 N to lift it, the efficiency of the pulley system is ......

ELECTRICAL SCIENCE

1. The charges that flow in solid conductors are called e........ and they are attracted towards a p....... p........ of supply. C..... and a........ are good conductors of e.........., whereas p........ c....... and b....... are good insulators.

2. The source of energy required to cause c...... flow is called e............ f.... and for this to happen a c...... needs to be created, with a 'g.' and 'r.....' conductor of low o.... resistance.

3. A well known formula for finding circuit resistance is R = ... which is based on O... L.. This states that the current flowing in a circuit is p.......... to the p........ d......... across the circuit.

A conductor's resistance depends on l. . . . ., c. . . . s. . . . . a. . . and r. . . . . The latter depends on the type of conductor m. . . . . . . used for the circuit conductors.

4. Resistors can be connected in s. . . . . or p. . . . . . ., or sometimes a combination of these methods is used. When resistors are connected in s. . . . ., resistance i. . . . . and circuit current d. . . . . In p. . . . . . . the resistance d. . . . . and more current flows.

5. Other circuit components capable of limiting current flow are i. . . . . and c. . . . . They possess r. . . . . measured in o. . .

6. When current flows through a d.c. circuit, resistance is the main opposition but in a.c. circuits it is i. . . . . Its symbol is . . and it is found from the ratio v. . . . . ./c. . . . . .

7. On a.c. supplies where mixed c. . . . . are used, a phase shift between current and voltage may occur and because of this a term known as p. . . . f. . . . . is used. If it is less than u. . . ., more current will be taken from the supply. For example, a 3 kW/240 V a.c. load with p.f. unity will take a current of 12.5 A, but if its power factor is 0.25, the current will be . . .

8. One method of indicating the north-seeking pole of a solenoid is by using the g. . . rule, where the f. . . . . . indicate current d. . . . . in the coil and the t. . . . the m. . . . . . . f. . . . around the pole.
A rule for finding i. . . . . . e.m.f. is called F. . . . . . . r. . . . h. . . rule and is particularly useful when dealing with e. . . . . induction. A single-loop a.c. generator operates on this principle, and is connected through s. . . r. . . . and rotated between the p. . . . of a magnet.

9. Transformers transform energy at one v. . . . . . and deliver it at another. Double-wound transformers have a supply winding called the p. . . . . . and a load winding called the s. . . . . The transformer's core is l. . . . . in order to reduce heating effects caused by e. . . c. . . . . . . When a transformer carries a load current, its iron losses remain c. . . . . . . at all loads. If a transformer has fewer turns on its secondary side than its primary side it is called a s. . . d. . . transformer.

10. A component which uses the c. . . . . . . effects of current is called a c. . . , b. . . . . . or accumulator. An accumulator consists of a number of c. . . . joined together in s. . . . . to create a large voltage. Two types commonly used are the l. . . a. . . and a. . . . . . . cells. The former type uses pure s. . . . . a. . . as an electrolyte, the latter uses potassium hydroxide in d. . . . . w. . . . The method of testing a b. . . . . . or accumulator for its s. . . . . . . g. . . . . . is to use a device called a h. . . . . An accumulator's p.d. is less than its e.m.f. because of its i. . . . . . . r. . . . . The ratio Ah on discharge to Ah on charge is called the Ah e. . . . . and is around 90% for most l. . . a. . . c. . . .

BASIC ELECTRONICS

1. A resistor is a p. . . . . . component whose chief property is r. . . . . In electronics, two methods of finding its value are used, (i) by c. . . . . c. . . and (ii) by p. . . . . . c. . . The former uses different c. . . . . b. . . . with the first two b. . . . indicating n. . . . . ., the third indicating a multiplier of n. . . . . . and a fourth b. . . indicating t. . . . . . . The latter method uses a mix of n. . . . . . and l. . . . . . stamped on the component.

2. A capacitor is a component that has the ability to store e. . . . . . . c. . . . . There are many different types available. The medium that separates one plate from the other is called the d. . . . . One type, which has to be connected correctly, is called a p. . . . . capacitor, for example, an e. . . . . capacitor. Other types, which are n. . . . ., include m. . . , c. . . . . . and p. . . . . capacitors. Some types of capacitor are identified using the methods described for resistors, but their values will be some submultiple of the f. . . . such as m. . . . . or p. . . . .

3. Another p. . . . . . component found in circuits is called an inductor. These are often used in electronics for producing high f. . . . . such as found in radio t. . . . . There are several types, with different c. . . construction, such as a . . c. . . , i. . . c. . . and f. . . . . . c. . .

An inductor is essentially a c. . . or w. . . . . . and on a.c. supplies particularly, it possesses i. . . . . (i.e. magnetic flux linkages creating an internal induced e.m.f.) which has the effect of l. . . . . . . current flow.

4. A diode is a s. . . . . device which acts as an electronic s. . . . . on d.c. supplies and a r. . . . . e. . . . . . on a.c. supplies. On d.c. it can only conduct in its f. . . . . . d. . . . . but on a.c. it passes p. . . . . . . h. . . p. . . . . A similar device, called a z. . . . diode, is used as a voltage s. . . . ., and this device operates in its r. . . . . . d. . . . . in its breakdown region. Other types of diode are called l. . . . e. . . . . . . d. . . . . , p. . . . . and d. . . . All the devices have two connections, called the a. . . . and c. . . . . .

5. Another type of semiconducting device is called a t. . . . . and these are frequently used as a switch and an a. . . . . It has three connections. A bipolar device made from npn or pnp material has connections called the b. . . , c. . . . . and e. . . . . . , while an FET unipolar device has connections called the s. . . . . , d. . . . and g. . .

6. One device which acts similarly to the diode is called a t. . . . . and this also has three connections. One of these is called the g. . . , which needs to be give a p. . . . . . . p. . . . before the device can conduct in the forward direction. It is made from s. . . . . . and is often referred to as a s. . . . . . c. . . . . r. . . . . (SCR). A modification of this device is called a t. . . . , which is two t. . . . . connected in opposite directions in order to provide f. . . w. . . r. . . . . of a.c.

7. A temperature-sensitive semiconductor resistor is called a t. . . . . This is a device designed for operation at both p. . . . . . . and n. . . . . . . temperatures.

8. In the electronics industry, small miniaturised packages involving i. . . . . c. . . . . . . are frequently used. They may be required to produced either d. . . . . . or a a. . . . . . . outputs and they are also used widely in many types of counting, visual displays and a. . . . . circuits.

9. The f. . . . . produced from the public supply sends current and voltage along a particular waveform which is s. . . . . in nature and known as a f. . . . . sine wave. This f. . . . . is at . . Hz. The time taken for one cycle (1 Hz) to be completed is called the p. . . . . . . time and for the public supply it is . . . . seconds.

In electronics and radio work other f. . . . . are also used, for example, sound waves at a. . . . f. . . . . (a.f.) are produced at 20 kHz–20 Hz and r. . . . f. . . . . (r.f.) start at about 20 kHz.

10. There are numerous types of waveform other than the f. . . . . sine wave and these are produced mostly from h. . . . . A s. . . . . wave is produced from a mixture of odd harmonic waves with smaller a. . . . . than the fundamental wave.

Answers 3

EXERCISE 1

1. a) 350 mΩ
   b) 0.75 J
   c) 0.255 MW
   d) 400 kV
   e) 22 μF

2. a) η For comparing output/input in the same units
   b) π The constant for any circle
   c) ρ A material's specific resistance
   d) $\propto$ Two quantities varying by the same amount
   e) $\geqslant$ A quantity may either have the same value as another quantity or alternatively be a higher value.

3. a) $d = \sqrt{(4A/\pi)}$
   b) $d = 1.78$ mm

4. a) 10^{11}
   b) 10^{-2}

5. a) $V_s = 250$ V
   b) $\cos\phi = 150/250 = 0.6$
   $\phi = 53°$

6. a) 6 cm^2
   b) 180°

7. $V = 300\pi\,(3^2 - 2.4^2)/4$
   $= 763.5$ cm^3

8. Draw a straight line of best fit near the point plotted. Create a right-angled triangle on the slope of the line. The y-axis represents a small increment of voltage and the x-axis a small increment of current. Since $R = V/I$, then the approximate value of $R = 2.35\ \Omega$

9. On your sketched half sine wave, draw and measure six mid-ordinates. The maximum value of the sine wave is to be considered unity and therefore no mid-ordinate must exceed 1. For the average value of induced e.m.f.
   $E_{\text{ave}} = (e_1 + e_2 + \ldots + e_6)/6 = 0.637$
   $E_{\text{r.m.s.}} = \sqrt{[(e_1^2 + e_2^2 + \ldots + e_6^2)/6]} = 0.707$
   Note: Answers will only be approximate.

10. The ratio $V_p/V_s = 20$
    therefore $V_s = 6600/20 = 330$ V

EXERCISE 2.1

1. 49.05 N

3. a) 7 cm b) 10.2 cm c) 125 N

4. $F = ma = 2 \times 12 = 24$ N

5. 150 N

6. $a = F/m = 40/400 = 0.1$ m/s^2

7. 98.1 J

8. 6 N/mm^2

EXERCISE 2.2

1. $\rho = m/V = 44\,000/4 = 11\,000$ kg/m^3

2. $m = \rho \times V = 1.3 \times 60 = 78$ kg

3. $V = m/\rho = 40\,000/1000 = 40$ m^3
   Since $V = l \times b \times h$ then $h = V/(l \times b) = 40/20 = 2$ m

4. Since displacement $V = 60 - 50 = 10$ cm^3 and $m = 30$ g then $\rho = m/V = 30/10 = 3$ g/cm^3 or 3000 kg/m^3.
   The heater has a mass of 0.03 kg

5. From formula [2.8]
   $\rho = m/V$
   $m = \rho \times V = 1.3 \times (2.5 \times 4 \times 20)$
   $= 260$ kg

EXERCISE 2.3

1. Since

$$F = mg = 15 \times 9.81 = 147.15 \text{ N}$$

and

$$W = Fd = 147.15 \times 10 = 1471.5 \text{ J}$$

then

$$P = W/t = 1471.5/30 = 49.05 \text{ W}$$

2. a) The forces are balanced and equal, hence frictional force = 12 N

b) When the applied force is 30 N, the frictional force is still 12 N
Thus, the resultant unbalanced force is $30 - 12 = 18$ N

c) From formula [2.5] $F = ma$ and $a = F/m = 18/4 = 4.5 \text{ m/s}^2$

3. a) $F = mg = 100 \times 9.81 = 981$ N
b) $W = Fd = 981 \times 3 = 2943$ J
c) $T = Fr = 981 \times 0.06 = 58.86$ Nm

4. Rope $ac = 225$ N and rope $bc = 270$ N

5. Here

$$F_1D_1 = F_2D_2$$
$$F_1 = F_2D_2/D_1$$

Thus

$$F_1 = 144/12 = 12 \text{ N}$$

6. a) Volume $V = Ah = 0.28 \times 0.8 = 0.224 \text{ m}^3$
From formula [2.8]

$$\text{Density} = \text{Mass/Volume}$$

Therefore

$$\text{Mass} = \text{Density} \times \text{Volume} = 1000 \times 0.224 = 224 \text{ kg}$$

b) From formula [2.11]

$$Q = mc \times \text{Temperature rise} = 224 \times 4200 \times 30 = 28.224 \text{ MJ}$$

7. From formula [2.20]

$$V_1/T_1 = V_2/T_2$$

where $V_1 = 9 \text{ m}^3$ and $V_2 = 5 \text{ m}^3$
In terms of absolute temperature

$$T_1 = 60 + 273 = 333 \text{ °C}$$

By transposition of formula

$$T_2 = (V_2 \times T_1)/V_1 = (5 \times 333)/9 = 185 \text{ °C}$$

Therefore the temperature has cooled to $185 - 273 = -88$ °C

8. By transposition of formula [2.10]
Expansion = Original length × Coeff. of expansion × Temp. rise
For aluminium, the expansion is
$15 \times 0.000\,026 \times 180 = 0.07$ cm

Its length has now increased to 15.07 cm

For copper, the expansion is
$15 \times 0.000\,017 \times 180 = 0.046$ cm

Its length has now increased to 15.046 cm

For glass, the expansion is
$15 \times 0.000\,009 \times 180 = 0.024$ cm

Its length has now increased to 15.024 cm

9. a) From formula [2.23]

$$P = 2\pi nT$$
$$T = P/2\pi n = 50\,000/(2 \times 3.142 \times 25) = 318.27 \text{ Nm}$$

b) A machine that is 85% efficient has losses of 15%. These losses occur as heat from the machine's windings (copper losses) and from its moving parts (rotational losses). The rotational losses are divided into iron losses (eddy currents and hysteresis losses) and mechanical losses (friction, bearings and windage).

10. In this question

$$\text{VR} = 4$$

and from formula [2.27] by transposition

$$\text{MA} = \text{Efficiency} \times \text{VR} = 0.75 \times 4 = 30$$

Hence

$$\text{MA} = \text{Load/Effort}$$

so

$$\text{Effort} = \text{Load/MA} = 300/30 = 10 \text{ N}$$

EXERCISE 3

1. a) $R = 44\ \Omega$, $P = V^2/R = (220 \times 220)/44 = 1100$ W
b) $R = 4\ \Omega$, $P = V^2/R = (220 \times 220)/4 = 12\,100$ W

2. $R = V/I = 240/5 = 48\ \Omega$
From formula [3.3], since $R = \rho l/A$ then $\rho = RA/l$
Hence $\rho = (48 \times 10 \times 10^{-6})/10 = 48\ \mu\Omega\text{m}$

3. Since
$\alpha = (R_t - R_0)/R_0 t$
then
$\alpha = (330 - 300)/(300 \times 25) = 0.004\ °C^{-1}$

4. An alternating current supply is one that is produced by an a.c. generator. It is cyclic in nature, passing through positive and negative frequency peaks. The public supply system generates 50 Hz or cycles every second.

A direct current supply is unidirectional and is not cyclic like a.c. It does not have a frequency and whilst it can be produced by a d.c. generator it is also produced chemically by batteries or by solar cells.

A transformer can only operate from an a.c. supply, since its operation principle is based on electromagnetic induction.

A rectifier is a device for converting a.c. into d.c. and therefore requires an a.c. supply.

5. a) $V = I \times R = 60 \times 0.06 = 3.6$ V
b) $P = V^2/R = (3.6 \times 3.6)/0.06 = 216$ W
c) $W = Pt = 216 \times 24 = 5.184$ kWh

6. From formula [3.14] $P = V \times I \times$ p.f.
a) By transposition, $I = P/(V \times \text{p.f.})$
hence $I = 45/(240 \times 0.4) = 0.468$ A
$I = 45/(240 \times 1) = 0.187$ A
A capacitor is used to improve the circuit power factor.

7. From formula [3.15]
% Efficiency = (Output/Input) × 100
By transposition
Input = (Output × 100)/% Efficiency
For the hoist,
Input = 300/70 = 4.285 kW
Since motor and hoist will be on the same shaft, the hoist's input is the motor's output.
Therefore, the motor's input is:
Input = 4.285/0.8 = 5.357 kW

8. See the diagram.

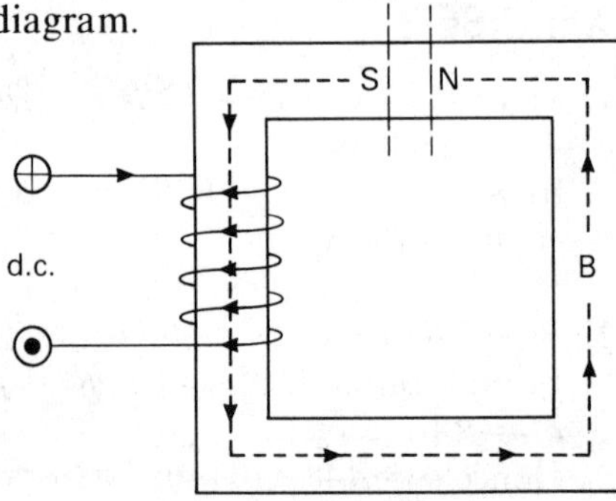

Another similar winding carrying equal flux would have the effect of neutralising the first flux and no circulation would occur. Cutting the core at the dotted lines produces a north pole where the flux leaves and a south pole where it enters.

9. a) $I = 0.961$ A; $V = 9.61$ V
b) $I = 0.140$ A; $V = 1.49$ V
c) $I = 0.43$ A; $V = 4.3$ V

10. See Figure 3.33 on page 49.

EXERCISE 4

1. a) brown, green, blue and silver; 16.5 MΩ and 13.5 MΩ
b) 470 pF or 0.47 μF; tolerance is ±20%, voltage 250 V

2. See the relevant sections in the text.

3. See the relevant sections in the text.

4. a) See the relevant sections in the text and the diagram.

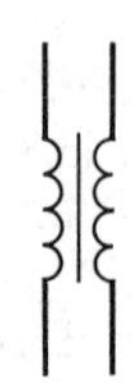

b) A transformer steps up or steps down voltage and current values.
The core is laminated to reduce the heat losses caused by eddy currents.

5. a) See page 60.
b) Both lamps will be dim because the voltage of the supply across each lamp is only 1.5 V. The diode does not have any influence on the circuit.
c) With the diode connection reversed only L_1 will light and be bright.

6. See page 63.

7. See pages 64–5.

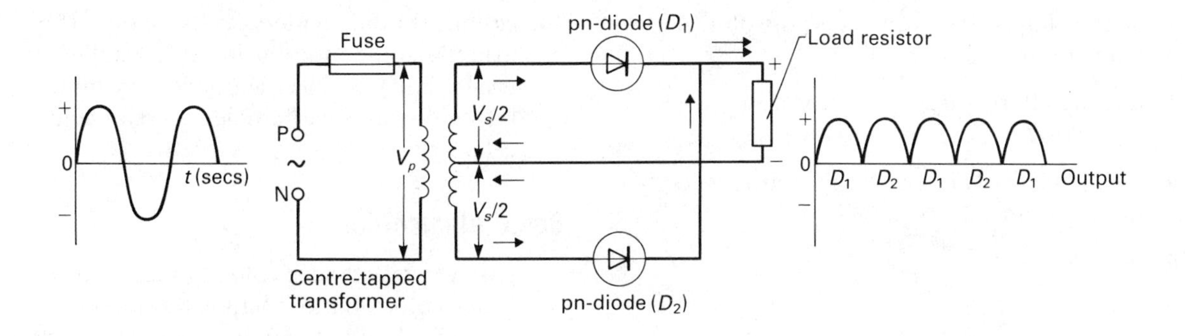

8. See the diagram, which shows full-wave rectification using a centre-tapped transformer and two diodes.

9. a) 141.44 V and 90 V
   b) See page 68.

10. a) 0 A and 20 A
    b) 60 Hz
    c) Instantaneous points
    d) $I = I_{max} \times \sin \phi = 10 \times \sin 120° = 8.66$ A

ANSWERS TO MULTIPLE CHOICE QUESTIONS

1. B	**6.** B	**11.** B	**16.** A
2. D	**7.** A	**12.** C	**17.** B
3. C	**8.** A	**13.** D	**18.** A
4. C	**9.** C	**14.** A	**19.** C
5. A	**10.** B	**15.** D	**20.** C
21. D	**26.** D	**31.** B	**36.** D
22. C	**27.** A	**32.** C	**37.** D
23. B	**28.** A	**33.** B	**38.** D
24. A	**29.** C	**34.** D	**39.** A
25. B	**30.** B	**35.** C	**40.** B
41. C	**46.** D	**51.** C	**56.** C
42. C	**47.** C	**52.** B	**57.** B
43. B	**48.** B	**53.** A	**58.** A
44. A	**49.** B	**54.** C	**59.** C
45. A	**50.** B	**55.** D	**60.** D
61. D	**66.** B	**71.** C	**76.** B
62. A	**67.** D	**72.** A	**77.** B
63. C	**68.** D	**73.** B	**78.** A
64. A	**69.** A	**74.** D	**79.** B
65. B	**70.** C	**75.** A	**80.** C
81. D	**86.** C	**91.** A	**96.** D
82. C	**87.** B	**92.** A	**97.** A
83. D	**88.** D	**93.** C	**98.** D
84. B	**89.** A	**94.** C	**99.** B
85. A	**90.** B	**95.** D	**100.** B

ANSWERS TO SHORT-ANSWER QUESTIONS

Terminology

1. metre; kilogram; watt; power; derived
2. prefixes; kilo; micro; 100 μF; 20 MΩ
3. 10^6; 10^{-2}
4. transposition; minus; denominator; quotient; squared; 9^2
5. multiplying; $2a^2 - 4ab$; $a^2 - b^2$
6. constant; circumference/diameter; πr^2; $V = Ah$
7. hypotenuse; sum; squares; opposite side
8. 90; 180; 360; right angles; diagonal; 1.414
9. instantaneous; root mean square; 0.707
10. proportional; inversely proportional; four; four

Mechanical science

1. mass; gram; kilogram; newtons; inertial; tensile; gravitational
2. weight; 9.81; acceleration; free fall; 50 N
3. magnitude; resultant; parallelogram
4. equilibrium; centre; gravity; principle; moments; clockwise moment; anticlockwise moment
5. mass; volume; temperature; ρ; 1000; float; sink; relative; 117 kg
6. energy; hotness; coldness; degrees; mercury; alcohol; Celsius; 100 °C; Fahrenheit; 212 °F; Kelvin; thermodynamic temperature; −273

7. acting; large; 100 N/m^2; gas; absolute temperature

8. turning effort; distance; newton metre; $P = 2\pi nT$

9. mechanical advantage; easy; hard; velocity ratio; distances

10. 66.66%

Electrical science

1. electrons; positive potential; copper; aluminium; electricity; polyvinyl chloride; bakelite

2. current; electromotive force; circuit; go; return; ohmic

3. *V/I*; Ohm's Law; proportional; potential difference; length; cross-sectional area; resistivity; material

4. series; parallel; series; increases; decreases; parallel; decreases

5. inductors; capacitors; reactance; ohms

6. impedance; Z; voltage/current

7. components; power factor; unity; 50 A

8. grip; fingers; direction; thumb; magnetic field; induced; Fleming's right hand; electromagnetic; slip rings; poles

9. voltage; primary; secondary; laminated; eddy currents; constant; step down

10. chemical; cell; battery; cells; series; lead acid; alkaline; sulphuric acid; distilled water; battery; specific gravity; hydrometer; internal resistance; efficiency; lead acid cells

Basic electronics

1. passive; resistance; colour code; printed code; colour bands; bands numbers; noughts; band; tolerance; numbers; letters

2. electric charge; dielectric; polarised; electrolytic; non-polarised; mica; ceramic; polyester; farad; microfarads; picofarads

3. passive; frequencies; tuners; core; air core; iron core; ferrite core; coil; winding; inductance; limiting

4. semiconducting; switch; rectifier element; forward direction; positive half pulses; zener; stabiliser; reverse direction; light emitting diodes; photodiodes; diacs; anode; cathode

5. transistor; amplifier; base; collector; emitter; source; drain; gate

6. thyristor; gate; positive pulse; silicon; silicon controlled rectifier; triac; thyristors; full wave rectification

7. thermistor; positive; negative

8. integrated circuits; digital; analogue; amplification

9. frequency; sinusoidal; fundamental; frequency; 50; periodic; 0.02; frequencies; audio frequency; radio frequencies

10. fundamental; harmonics; square; amplitudes